T0330588

STATISTICAL PROCESS CONTROL FOR REAL-WORLD APPLICATIONS

STATISTICAL PROCESS CONTROL FOR REAL-WORLD APPLICATIONS

WILLIAM A. LEVINSON

CRC Press
Taylor & Francis Group
Boca Raton   London   New York

CRC Press is an imprint of the
Taylor & Francis Group, an **informa** business

CRC Press
Taylor & Francis Group
6000 Broken Sound Parkway NW, Suite 300
Boca Raton, FL 33487-2742

© 2011 by Taylor & Francis Group, LLC
CRC Press is an imprint of Taylor & Francis Group, an Informa business

No claim to original U.S. Government works

Printed on acid-free paper
Version Date: 20141204

International Standard Book Number-13: 978-1-4398-2000-1 (Hardback)

Library of Congress Cataloging-in-Publication Data

Levinson, William A., 1957-
  Statistical process control for real-world applications / William A. Levinson.
     p. cm.
  Includes bibliographical references and index.
  ISBN 978-1-4398-2000-1 (hardback)
     1. Process control--Statistical methods. 2. Quality control--Statistical methods. I. Title.

TS156.8.L484 2010
658.5--dc22   2010042056

Visit the Taylor & Francis Web site at
http://www.taylorandfrancis.com

and the CRC Press Web site at
http://www.crcpress.com

Contents

Preface: Why This Book?

Most books on statistical process control (SPC) rely on the assumption that measurements from manufacturing processes follow the normal distribution or bell curve—an assumption that often does not hold up on the factory floor. This book equips the reader to deal with nonnormal applications scientifically, and also to explain the methodology to suppliers, customers, and quality system auditors. It also applies this maxim from Robert A. Heinlein's *Starship Troopers* (1959) for deployment of the advanced statistical methods to the factory floor:

> If you load a mud foot down with a lot of gadgets that he has to watch, somebody a lot more simply equipped—say with a stone ax—will sneak up and bash his head in while he is trying to read a vernier.

The *vernier* part of the job includes sophisticated statistical techniques that the quality engineer or industrial statistician uses offline to set up control charts and calculate process performance indices for nonnormal applications. The same applies to mathematical derivations and goodness-of-fit tests to satisfy the practitioner, customers, and quality system auditors that the techniques are in fact valid. The *stone ax* part covers the need for the frontline production worker to enter his or her measurements and get immediate and automatic feedback on whether the sample statistics are inside or outside the control limits. Any time that workers must spend on data entry, not to mention calculations or detailed interpretation of control charts, is time they cannot spend on actual production that pays wages and profits.

A spreadsheet can often perform all the necessary calculations and then give the operator an immediate signal (such as changing the background color of a cell) if a point is outside the control limits. The spreadsheet becomes a *visual control* that makes the status of the process obvious without the need to interpret numbers or even a graph. This combines extremely accurate and relevant calculations (the vernier) with the simplicity of a go/no-go signal (the stone ax). This book shows how to deploy these methods with Microsoft Excel, and the methods can in some cases be adapted to Corel Quattro Pro or IBM Lotus.[1] The factory may already use a commercial SPC program; if so, many of these programs will accept user-defined control limits.

This book will address the following key issues.

1. Application of the normal distribution to processes that actually follow the gamma, Weibull, or other nonnormal distribution can result in very inaccurate statistical process control charts and misleading process capability indices. This book shows how to calculate appropriate control limits and performance indices for nonnormal distributions. It

also improves on the traditional charts for sample range and standard deviation by showing how to set up R (range) and s (standard deviation) charts with known and exact false alarm limits, and it shows how to construct range charts for nonnormal distributions.

2. Davis and Kaminsky (1990; see also Burke, Davis, and Kaminsky, 1991) use the phrase "statistical terrorism" to describe dysfunctional relationships between large companies and suppliers, and this issue applies especially to process capability and performance indices. The large company may dictate the use of inappropriate statistical methods or sampling plans that deliver inaccurate results. The following considerations are often overlooked:

   - The central limit theorem applies to the averages of large samples but not to individual measurements that are either in or out of specification. It therefore cannot be invoked for calculation of capability or performance indices. If the process does not follow the normal distribution, it is necessary to fit the appropriate nonnormal one and then calculate the nonconforming fraction. The Automotive Industry Action Group (2005) sanctions this approach, so it should be acceptable to International Organization for Standardization ISO/TS 16949 auditors.

   - The historical database must be adequate, and capability indices that are based on ten or twenty measurements are almost meaningless.

   - The process must be in a state of statistical control.

3. Batch-and-queue processing is a common obstacle to just-in-time (JIT) manufacturing (Levinson and Rerick, 2002, 130–138). Nested variation sources also complicate SPC and process capability calculations. This book shows how to set up appropriate control charts and calculate accurate process performance indices.

4. Imperfect gage capability (reproducibility and repeatability) affects the power of SPC charts to detect process shifts, and it also affects outgoing quality. This book shows how to deal quantitatively with this issue.

5. This book's primary focus is not on situations that involve correlated quality characteristics (multivariate systems), but the last chapter provides an overview and references to useful sources.

Endnotes

1. The Visual Basic for Applications functions on the user disk work in Microsoft Excel. Corel's WordPerfect Office 12 also has VBA capability and, according to http://www-01.ibm.com/support/docview.wss?uid=swg21110222, LotusScript is similar to VBA. We have not, however, worked with Corel Quattro Pro or Lotus 1-2-3.

About the Author

William A. Levinson, P.E., is the owner of Levinson Productivity Systems, P.C. which specializes in quality management, industrial statistics, and lean manufacturing. He is the author of *Henry Ford's Lean Vision: Enduring Principles from the First Ford Motor Plant* (Productivity Press, 2000) and *Beyond the Theory of Constraints* (Productivity Press, 2007), as well as several books on quality and statistical process control (ASQ Quality Press). He holds professional certifications from the American Society for Quality (ASQ), Society of Manufacturing Engineers, APICS, and the Association for Operations Management, and he is a Fellow of ASQ.

Supplementary Resources Disclaimer

Additional resources were previously made available for this title on CD. However, as CD has become a less accessible format, all resources have been moved to a more convenient online download option.

You can find these resources available here: https://www.routledge.com/9781439820001

Please note: Where this title mentions the associated disc, please use the downloadable resources instead.

Introduction

What to Expect From This Book

This book specializes in three topics: nonnormal distributions, nested normal systems, and confidence limits for capability indices. Although it is not designed as a primary textbook, it includes some exercises to allow the reader to apply its content to simulated process data. The accompanying CD-ROM includes data for the exercises (Exercises.xls), so the user does not have to type the data in manually.

The book also improves on the traditional approaches to the charts for sample range (*R*), sample standard deviation (*s*), and attributes (*np*, *p*, *c*, and *u*). The traditional approaches all treat these statistics as if they follow a normal distribution, which they do not. The normality assumption improves with increasing sample size for *R* and *s*, and with higher nonconformance or defect rates for the attribute charts, but the latter condition is highly undesirable. Modern computational methods, most of which are easily deployable in a spreadsheet, can calculate exact control limits for all these applications in real time.

The book begins with an overview of traditional statistical process control, including goodness-of-fit tests that might reveal whether the traditional methods are appropriate. Then it addresses the issue of normal versus nonnormal distributions.

Normal versus Nonnormal Distributions

The normal distribution's probability density function (pdf) is

$$f(x) = \frac{1}{\sqrt{2\pi}\sigma} \exp\left(-\frac{1}{2\sigma^2}(x-\mu)^2\right)$$

where μ is the process mean and σ^2 is the variance. Typical shorthand for such a distribution is $N(\mu,\sigma^2)$, with the one in Figure 0.1 being $N(100,4)$. The standard deviation, $\sigma = 2$, is the square root of the variance. The ordinate $f(x)$ is the relative chance of getting a particular number from this distribution.[1]

Since the standard deviation is 2, the traditional Shewhart 3-sigma control limits are $100 \pm 3 \times 2$, or [94,106]. We expect a false alarm rate of 0.00135 at each control limit, and we will compare this to the actual situation below.

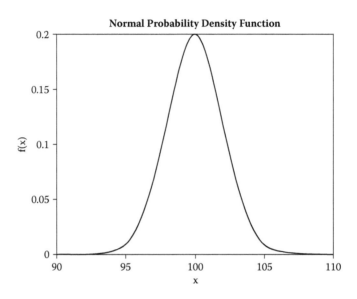

FIGURE 0.1
Normal (Gaussian, bell curve) distribution.

Figure 0.1 is often wishful thinking in real factories. Processes often look more like Figure 0.2, and it is not because they are out of control; it is their natural behavior. As stated by Jacobs (1990, 19–20),

> One of the major sources of frustration in the application of statistical process control (SPC) methods to a chemical process is the prevalence of variables whose values have distributions that are, by nature, distinctly nonnormal. Typically, methods used to analyze these variables are based on the normal distribution and, as such, are unrealistic.

This reference cites quality characteristics, such as moisture content and impurities, with a natural limit of zero as examples. Our experience is that impurities follow the gamma distribution, which is the continuous scale analogue of the Poisson distribution—the standard model for undesirable random arrivals such as defects. The reference adds that distributions for characteristics near a physical limit, such as tensile strength, saturation (the coexistence of the liquid and vapor phases in a boiling liquid), and phase changes (such as melting or boiling), are often nonnormal.

In the case of a brittle material, failure at any point generally creates a stress concentrator, cracks that propagate rapidly, and fracture. Kapur and Lamberson (1977, 77–78) state that the extreme value distribution is a reasonable model for materials that fail at their weakest points. If the failure strength depends on the average of strong and weak elements, as might be the case for a ductile material, the distribution should be normal. This is in

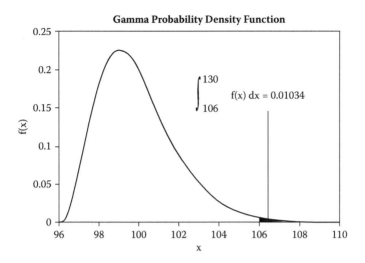

FIGURE 0.2
Gamma distribution.

fact true of the ultimate tensile and yield strength of steel. The Weibull distribution applies to other material characteristics, and factors such as heat treatment and surface finish may change the distribution function. The need to identify the correct distribution therefore cannot be overemphasized.

A one-sided specification is a sign that the process's distribution *might* be nonnormal. As an example, customers rarely set lower specification limits for impurity levels in chemicals. Semiconductor manufacturers don't care how few particles show up in process equipment. There will only be an upper specification limit (USL), and there is usually only a lower specification limit for quality characteristics such as tensile strength. In all these cases, the quality characteristic cannot be less than zero. If a significant portion of the traditional bell curve extends below zero, it is clearly not a suitable model for the process.

As stated above, impurities and particle counts often follow the gamma distribution. Figure 0.2 shows a gamma distribution which, like the normal distribution in Figure 0.1, has a mean of 100 and a variance of 4. The integral is from 106, the Shewhart 3-sigma upper control limit, to infinity. The chance of getting a point outside the upper control limit of 106 is 0.01034—*more than seven times what the normal distribution predicts.* Shop personnel might accept a false alarm risk of 0.00135, or 0.0027 for a two-sided specification, but they are likely to take a dim view of a process control that forces them to use an out-of-control action procedure for one out of every hundred samples. Aesop's fable about the boy who cried "Wolf!" too many times is highly instructive.

Figure 0.3 shows an individuals chart (X chart) as generated by Minitab for 100 measurements from this distribution. Clearly, $\mu + 3\sigma = 106$ (or, in this case, 105.3, by average moving range) is not an appropriate upper control limit for this process.

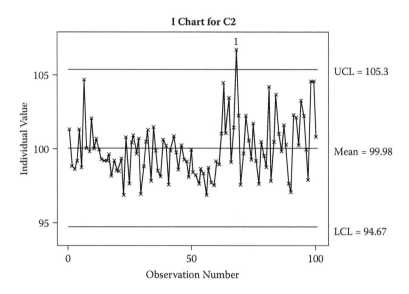

FIGURE 0.3
Behavior of a gamma distribution on an X (individuals) chart.

This book shows how to calculate *Shewhart-equivalent control limits* for non-normal processes; that is, limits with false alarm rates of 0.00135 at each end. (The user can select other sensitivities if desired.) Nonnormal distributions are not, however, the only ones that complicate SPC.

Nested Normal Distributions

Figure 0.4 is a Minitab-generated control chart of 50 subgroups of 4 measurements. The s chart for process variation is in control, but the x-bar chart for process mean has points outside *both* control limits. This suggests that the control limits have not been calculated properly.

The problem lies in the selection of the *rational subgroup*—one that represents a homogenous set of process conditions. In this example, however, the 50 so-called subgroups of 4 came from a batch process. That is, four measurements were taken from each oven-load of parts or kettle-full of material (Figure 0.5).

Process conditions are often *not* homogenous from one batch to the next. There is both between-batch and within-batch variation, and between-batch variation is likely to be the larger quantity. If the x-bar chart's control limits account for only the within-batch variation, they will be too tight; hence the points outside both control limits. The control chart for variation will meanwhile remain in control provided that within-batch variation remains constant. If any of the reader's control charts look like Figure 0.4, this book shows how to handle the situation.

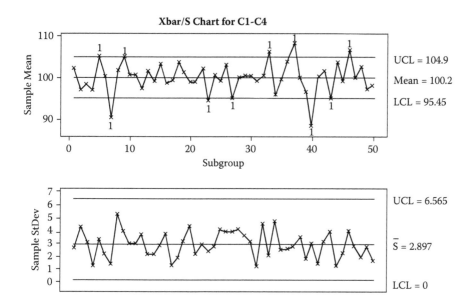

FIGURE 0.4
What's wrong with this picture?

Meanwhile, process capability and process performance indices are among the most important process metrics, and it is important to calculate them correctly. A capability or performance index that accounts only for within-batch variation will be far too optimistic.

Process Capability Indices

Process capability and performance indices measure a manufacturing process's ability to meet specifications. They are statements to the customer about

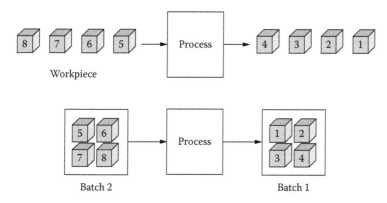

FIGURE 0.5
Single-unit processing (top) and batch processing (bottom).

the expected nonconformance rate or defects per million opportunities (in the absence of special or assignable cause variation). A Six Sigma process has six standard deviations between its nominal and its specification limits. If the process remains centered, the nonconformance rate due to random variation will be 2 parts per billion (ppb). If the process shifts 1.5 standard deviations (and any decent control chart will detect this quickly), the nonconformance rate will be 3.4 parts per million (ppm). Many customers require regular process capability reports from their suppliers, and adequate process capability is often a requirement of the quality management system.

There are several often-overlooked problems with capability indices. Even if the process follows the normal distribution, the estimated indices will be unreliable if the historical dataset is too small. Figure 0.6 shows the 95% (2-sided) confidence interval for the process performance index

$$P_p = \frac{USL - LSL}{6\hat{\sigma}_p} = 2.0,$$

where the standard deviation is estimated from the entire dataset as opposed to subgroup statistics. $P_p = 2.0$ is a Six Sigma process. The confidence interval for a sample of 10 is [1.10, 2.91], which means 95% confidence that the true P_p is somewhere between unacceptable (1.33 is the generally accepted minimum for a "capable" manufacturing process) and outstanding.

The author has seen attempts to report capability indices on the basis of 10 or even fewer measurements. Even when 100 measurements are available, the confidence interval will still be substantial—[1.72, 2.28] for the example under consideration.

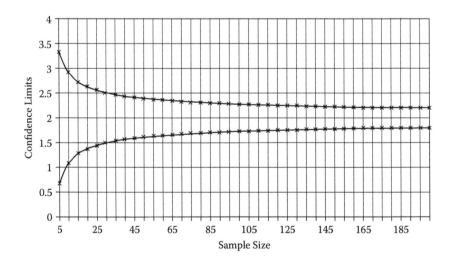

FIGURE 0.6
Ninety-five percent 2-sided confidence limits for Six Sigma performance.

If the process is nonnormal, there is no reliable relationship whatsoever between the traditional capability or performance index and the nonconforming fraction, aside from the fact that a higher index will always be better for the process in question. (If it is higher than that of a different process with a different distribution, however, there is no guarantee that it will deliver fewer nonconformances than the other process.) Consider the distribution in Figure 0.2 again, and assume that the upper specification limit is 112, six standard deviations above the mean. The statement $PPU = 2$, where

$$PPU = \frac{USL - \mu}{3\sigma_P},$$

suggests that the nonconforming fraction will be 1 part per billion at this specification limit *if the underlying distribution is normal.*

The actual distribution is, however, a gamma distribution;

$$f(x) = \frac{\gamma^\alpha}{\Gamma(\alpha)}(x - \delta)^{\alpha-1}\exp(-\gamma(x - \delta))$$

where α is the shape parameter, γ is the scale parameter, and δ is the threshold parameter. (For integers, $\Gamma(x) = (x - 1)!$, e.g., $\Gamma(4) = 3! = 6$.) δ is the *guarantee time* in reliability applications, and it is the customer's friend. It means that no part will fail before δ time periods (or other units of use) through random variation. In many SPC applications, however, it unfortunately means that the process will never deliver fewer than δ impurities or particles. The distribution has the following properties:

$$\mu = \frac{\alpha}{\gamma} + \delta \quad \text{and} \quad \sigma^2 = \frac{\alpha}{\gamma^2}.$$

Given $\alpha = 4$, $\gamma = 1$, and $\delta = 96$, $\mu = 4 + 96 = 100$ and $\sigma^2 = 4$; the nonconforming fraction above $USL = 112$ is:

$$1 - \int_{96}^{112} \frac{1}{\Gamma(4)}(x - 96)^3 \exp(-(x - 96))dx$$

$$= 1 - \int_0^{16} \frac{1}{6}x^3 e^{-x}\, dx$$

$$= 1 - \left[-e^{-x}\left(\frac{1}{6}x^3 + \frac{1}{2}x^2 + x + 1\right)\right]\Big|_0^{16}$$

$$= e^{-16}\left(\frac{1}{6}16^3 + \frac{1}{2}16^2 + 16 + 1\right) = 9.314 \times 10^{-5}$$

A nonconformance rate of 9.314×10^{-5} (93 defects per million opportunities) is *almost five orders of magnitude greater than the expected nonconformance rate of one per billion.* It is consistent not with Six Sigma process capability, but with *PPU* ≈ 1.25. This process is not even capable under the generally accepted minimum standard of 1.33. This is, of course, because of the distribution's long upper tail.

The statement that the true nonconformance rate is consistent with a 1.25 capability index suggests the concept of an *equivalent capability index*—the capability index of a normally distributed process that has the same nonconformance rate, and the Automotive Industry Action Group's SPC manual (2005) sanctions this approach. This book will show how to apply this concept.

Chapter Overview

Traditional Control Charts

Chapter 1 shows how to use traditional Shewhart control charts. It begins with a discussion of variation and accuracy, and then it treats false alarm risks, a concept that will apply to charts for nonnormal distributions. It also covers the vital issue of testing data for normality by using probability plots and the chi square goodness-of-fit test. Probability plotting and the chi square test can also test for goodness of fit to nonnormal distributions. The chapter content is as follows.

- Variation and accuracy
- Statistical hypothesis testing
  - Control charts test the hypothesis that the process is in control, i.e., that its variation and accuracy have not changed.
  - This section will also be very useful to readers who are taking courses in design of experiments (DOE) or are trying to use it in their workplaces.
- Standard control charts
  - x-bar (sample average) and R (sample range) charts
  - x-bar/s (sample standard deviation) charts
  - X (individual measurement) charts
  - z chart for sample standard normal deviate (useful when the process makes products with different nominals) and the corresponding χ^2 chart for variance.
  - Western Electric zone tests. The center line for a nonnormal distribution's chart should be the median (50th percentile) and not

the mean. The Zone C test for eight consecutive points above or below the center line relies on the fact that, if the process is in control, a point has a 50:50 chance of being on either side of the line. The mean is also the median for the normal distribution, but this is not true for the Weibull and gamma distributions.

- Process capability and process performance indices
- Tests for goodness of fit
- Multiple attribute control charts
  - This replacement for the traditional p (fraction nonconforming), np (number nonconforming), c (defect count), and u (defect density) chart is essentially a check sheet or tally sheet that can track multiple problem sources and provide a visual signal (change in spreadsheet cell color) for an out-of-control condition.
  - It uses the exact binomial or Poisson distribution to calculate its control limits as opposed to the normal approximation, so it does not require the expectation that four to six pieces in every sample will be defective or nonconforming.

Nonnormal Distributions

Chapter 2 introduces the gamma and Weibull distributions. It shows how to fit them to data, how to develop appropriate control charts, and how to compute process performance indices.

- Gamma distribution
  - How to fit data to the three-parameter gamma distribution
  - Tests for goodness of fit: quantile-quantile plot and chi square test
  - Control chart for the gamma distribution
  - Process performance indices for the gamma distribution
- Weibull distribution
  - How to fit data to the three-parameter Weibull distribution
  - Tests for goodness of fit, control charts, and performance indices are developed as for the gamma distribution.
- Left-censored distributions and lower detection limits
  - Right-censoring is common in reliability applications and life testing. Left-censoring takes place when the instrument that measures the quality characteristic has a lower detection limit.

Table 0.1 summarizes the available methods and control charts.

Table 0.2 summarizes the ability of software (as shown in the book) to perform the necessary tasks. Note that all the control charts can be

TABLE 0.1

Available Control Charts and Methods

Distribution	Control Charts				Confidence Limits for Process Performance Index
	X	x-bar	R[a]	S[a]	
Normal	Yes	Yes	Yes	Yes	Yes
Gamma	Yes	Yes	Yes	No	c
Weibull	Yes	No[b]	Yes	No	Yes

[a] With exact false alarm risks, plus methods to calculate power for any given process shift

[b] Not possible due to the nature of the distribution

[c] Lawless (1982, 218) provides the only example that we know in which an attempt was made to find the confidence limits for the survivor function, which corresponds to the process performance index, of a gamma distribution. We reproduced these results independently, but, as only a single test was possible, a caveat should be attached to any results that are reported to customers.

deployed on the factory floor on either a stand-alone spreadsheet (Excel, Quattro Pro, or Lotus) or a spreadsheet that can use Visual Basic for Applications (VBA) functions. *Exact limits* refers to sample standard deviation and range charts with false alarm risks of exactly 0.00135 as opposed to the traditional ones that use normal approximations for the underlying distributions. A blank does not necessarily mean that the software is not capable of the indicated task, but rather that we have not verified its usability for this purpose. The reader should go to the vendor's Web site or contact the vendor directly to determine whether the software meets a particular need.

The Environmental Protection Agency has meanwhile developed two free downloadable packages known as ProUCL (UCL stands for upper confidence limit) and SCOUT.[2] They perform many of the tasks that this book describes, and they devote considerable attention to the gamma distribution because of its ability to model contaminant levels. Both programs require Microsoft .NET Framework 1.1, which can be downloaded for free from Microsoft if it is not already installed on the computer.

Variation Charts for Nonnormal Distributions

Chapter 3 begins by making obsolete the traditional range and sample standard deviation charts for the normal distribution. These charts rely on 3-sigma approximations for the highly nonnormal distributions of the sample range and standard deviation, and this was probably the best available technology before the invention of high-speed computers. It is, however, possible to construct R and s charts with exact false alarm risks. The control limit for the R or s chart can, for example, be set to deliver a false alarm risk

TABLE 0.2

Software Capability

Application	Spread-sheet[b]	MS Excel VBA[c]	Minitab 15	StatGraphics Centurion	MathCAD
Normal Distribution					
X and x-bar	Yes	—	Yes	Yes	—
s chart, exact limits	Yes	—	—	—	—
R chart, exact limits	—	Yes	—	—	Yes
Confidence intervals for P_p, PPL/PPU, P_{pk}	P_p only	Yes	—	Yes[a]	Yes
Gamma Distribution					
X and x-bar	Yes	—	Yes	Yes	—
R chart, exact limits	—	Yes	—	—	—
Confidence interval, PPL/PPU	—	d	—	—	—
Fit uncensored	—	Yes	Yes	Yes	Yes
Fit left-censored	—	Yes	—	Yes	Yes[e]
Weibull Distribution					
X	Yes	—	—	—	—
x-bar		Not possible			
R chart, exact limits	—	Yes	—	—	—
Confidence interval, PPL/PPU	—	Yes	Use Weibull conf. limits	Graphical Weibull	Yes
Fit uncensored	—	Yes	Yes	Yes	Yes
Fit left-censored	Yes	Yes	Yes	Yes	Yes[e]
Binomial and Poisson Distributions (nonconformances and defects, respectively)					
Multiple attribute chart, exact limits	Yes	—	—	—	—

[a] Use one-sided normal distribution tolerance intervals to get confidence limits for *PPL* and *PPU*.

[b] Stand-alone spreadsheet functions; may need fitted parameters from other programs.

[c] Excel with Visual Basic for Application (VBA) functions on the user disk. The user may, in some cases, have to change bisection interval limits or Romberg integration parameters.

[d] There is very little in the literature about maximum likelihood estimation of the confidence interval for the cumulative density function or survivor function of the gamma distribution, but a VBA algorithm appears successful.

[e] Left-censored distributions apply to quality characteristics with lower detection limits, e.g., for impurities or trace materials. The issue of a lower detection limit, or "nondetect," is also of interest to the Environmental Protection Agency. The MathCAD algorithm required substantial user intervention to set appropriate bisection intervals for nested bisection routines.

of exactly 0.00135 for which the Type II risk of not detecting a given increase in process variance can be calculated. The chapter then uses the same methods to develop R charts for nonnormal distributions, and these also have known and exact false alarm risks.

Nested Normal Distributions

Chapter 4 treats batch processes that have nested variation sources—between-batch and within-batch sources. This includes calculation of the variance components and the process performance index.

Confidence Intervals for Process Performance Indices

Chapter 5 shows how to find confidence intervals for normal and Weibull process performance indices. Remember that the process performance index is a statement to the customer about the expected nonconforming fraction due to random variation, and a confidence limit is a quantitative estimate of the worst case. The chapter handles

- Confidence intervals for normal P_p
- Confidence intervals for normal *PPL* and *PPU*
  - The noncentral t distribution
  - One-sided tolerance intervals
- Confidence intervals for normal P_{pk}
  - The bivariate noncentral t distribution
  - Two-sided tolerance intervals
- Confidence intervals for Weibull and gamma *PPL* and *PPU*

The Effect of Gage Capability

Chapter 6 treats the effect of gage reproducibility and repeatability (GRR, or R&R) on control charts, process capability, and outgoing quality.

Multivariate Systems

Chapter 7 treats systems in which there are specifications for two or more *interdependent* process characteristics.

Try Minitab Free!

Minitab is the leading statistical software used to learn statistics and improve quality. Its intuitive interface and rich graphs make it the package of choice for thousands of businesses and more than 4,000 colleges and universities worldwide.

This book references Minitab. To see firsthand how easy it is to analyze your data, download a free 30-day trial at www.minitab.com/minitab-free-trial.

Software and User Disk

The current versions of Minitab® Statistical Software[3] and StatGraphics Centurion can perform most of the calculations that are necessary to handle nonnormal distributions, and MathCAD can be programmed to do so. The book contains examples from all three software packages, and the first two are designed explicitly to perform the necessary calculations and generate supporting material like probability plots and histograms.

This book also devotes considerable attention to spreadsheets. If commercial SPC software is not available—and any that is, must be programmable to accept user-defined control limits for nonnormal distributions—Excel and its relatives such as Corel Quattro Pro and IBM Lotus can accept manual data entry and give the production worker a visual signal, such as a change in cell color, when an out of control condition is present.

This leads to an important practical consideration for the offline data analysis itself. Although this book is not primarily about Microsoft Excel, nor does the author claim to be a software professional, the user disk includes Visual Basic for Applications functions to make the following tasks as routine as possible:

1. Fit nonnormal distributions to data by maximum likelihood optimization.
2. Create data columns that Excel's built-in scatter chart capability can make into a quantile-quantile plot for normal and nonnormal distributions.
3. Perform the chi square goodness-of-fit test for normal and nonnormal distributions.
4. Calculate cumulative distributions for sample ranges from the normal and nonnormal distributions.

License and Disclaimer, Visual Basic for Applications Functions

Levinson Productivity Systems, P.C. retains all rights to the VBA functions, and grants the book's purchaser a license to use them on a single computer for job-related, academic, evaluation, and any other purpose. (The functions may be stored on more than one computer, e.g., in the workplace and at home, as long as only one is in actual use.) The company cannot provide individualized technical support, but it will post updates and relevant information at its Web site.

These functions have been tested successfully at least once as demonstrated in the book, but they have not been tested against all conceivable applications. No warranties are therefore expressed or implied. A function that does the chi square test or quantile-quantile plot properly for one set of data *should* do it correctly for any set of data, and the same goes for the ability to fit the nonnormal distributions. In the latter case, however, the user must specify a maximum allowable shape parameter that exceeds all possibilities for the bisection procedure that the functions use, but not so large as to cause computational overflow.

The cumulative distributions for sample range use Romberg integrations for which the tolerances must be sufficiently small to ensure a correct answer, but not so small as to prevent convergence; in addition, the value of "infinity" must be specified correctly. The one for the range of the normal distribution was tested successfully against tabulated data, but those for the ranges of gamma and Weibull distributions could be tested only against results from MathCAD.

User Disk

The user disk contains the following material:

1. Simulated datasets for the examples in the book are in SIMULAT. XLS. The reader can experiment on them with whatever software his or her company uses.
2. Control chart spreadsheets are in SPC.XLS; the reader can use them as they are or modify them for use in his or her factory.
3. Datasets for the exercises in the book are in EXERCISES.XLS.
4. Visual Basic for Applications functions for Excel spreadsheets are in .bas files (for direct importation into spreadsheets) and .txt files (for copying and pasting into Excel's VBA editor). They were tested as shown in VBA.XLS. Our experience is that, if more than one function is in use on a spreadsheet (which requires more than one module to be present in the VBA editor), all of them are re-executed whenever one is updated. This is why only the test results, as opposed to live functions, are present in VBA.XLS; the Romberg integrations can take several minutes to fill in an entire table. Documentation of these functions is in the file VBA.DOC.
5. MathCAD files cited in the book are in the .MCD files. MathCAD 8.0 or higher (see http://www.ptc.com) is required for their use.

Endnotes

1. The chance of getting a specific number is a differential quantity.
2. http://www.epa.gov/esd/tsc/software.htm and http://www.epa.gov/esd/databases/scout/abstract.htm as of January 2010. Alternatively, do a Google search on site epa.gov and "ProUCL" or "SCOUT."
3. MINITAB® and all other trademarks and logos for the company's products and services are the exclusive property of Minitab Inc. All other marks referenced remain the property of their respective owners. See minitab.com for more information.

1

Traditional Control Charts

This chapter will review the traditional control charts that assume normality for the underlying process. It will also provide an overview of the central concepts of variation and accuracy. Many of the concepts in this chapter will carry over into the chapters on nonnormal and nested normal distributions.

Variation and Accuracy

A manufacturing process's ability to meet specifications depends on two factors: variation and accuracy. Variation is relative to the specification width, and the process capability and process performance indices measure this relationship. Accuracy means that the process mean is at the nominal.

A good way to teach these concepts is to treat the specification as a target, and the process as a firearm. The bull's-eye is the process's nominal measurement or desired point of aim. A shot that hits the target is in specification, and one that misses is out of specification; it is rework or scrap.

Figure 1.1 shows simulations of 250 shots from each of four firearms. The rifle, which puts spin on its projectile, produces tighter shot groups than the smoothbore musket. The rifle is *capable* of hitting the target consistently. If its sights are not centered on the bull's-eye (nominal), it is, however, more likely to miss. Under these conditions, the rifle is *not in control*. Adjustment of the rifle's sights can bring the aiming point back to where it belongs; that is, the process can be adjusted.

There is so much spread in the smoothbore musket's shot pattern that it can easily miss even when it is aimed straight at the bull's-eye. Even when the musket is *in control*, it is simply *not capable* of meeting specifications consistently. In the "in control but not capable" example, adjustment of the musket's sights will only make matters worse because the musket's current aiming point is the nominal.

A factory cannot adjust quality into a process that is not capable. Figure 1.2A shows the consequences of trying to do so, and it was generated as follows:

$$x_i = r_i \cos(\theta_i) - x_{i-1} \quad y_i = r_i \sin(\theta_i) - y_{i-1} \text{ where}$$

$$r_i \sim N(0,1) \quad \theta_i \text{ is random from 0 to } 360°, \text{ and } x_0 = y_0 = 0$$

In other words, if $(x_1, y_1) = (0.3, -.4)$, the sights are adjusted -0.3 (left) and $+0.4$ (up) in an attempt to improve the next shot.

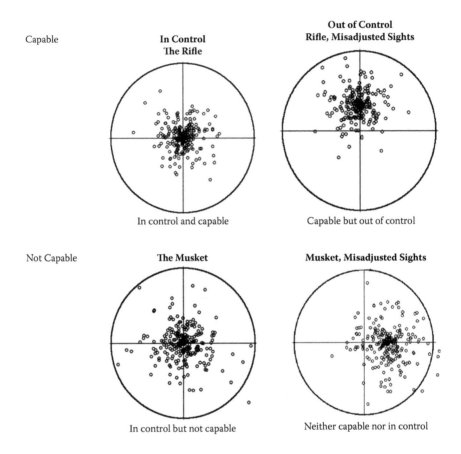

FIGURE 1.1
Variation and accuracy.

Figure 1.2B (Minitab) is the one-dimensional counterpart of Figure 1.2A, and it illustrates what overadjustment does to a factory process. The target or nominal is 100, and the actual variation is 1. The control algorithm sets the new aiming point to $100 + (100 - X)$ or simply $200 - X$ to "compensate" for the deviation from target; as an example, if $X = 95$, then the tool is adjusted to 105 in the hope that the next result will be on target.[1]

This example shows that efforts to compensate for common cause or random variation are worse than futile, and they make the process a lot worse. The consequences of *overadjustment* or *tampering* reinforce the adage, "If it ain't broke, don't fix it."

There is no inconsistency whatsoever between the admonition, "If it ain't broke, don't fix it," and the need for continuous improvement (kaizen). Fixing means correction of an *assignable cause* or *special cause* problem such as a misadjusted rifle sight. Improvement means reduction of *common cause* or *random* variation, for example, by replacement of the musket with a rifle.

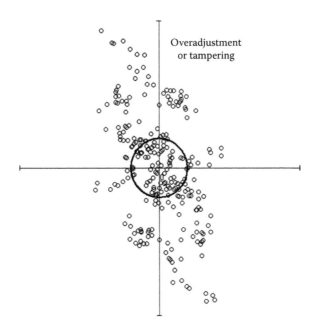

FIGURE 1.2A
Overadjustment or tampering.

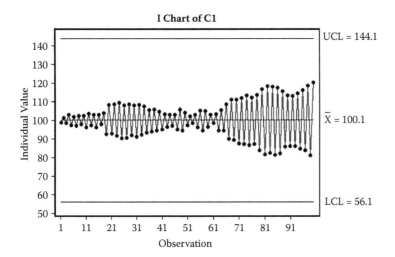

FIGURE 1.2B
Overadjustment: Autoregressive AR(1) model.

Furthermore, there are scientific ways to adjust a process on the basis of feedback from that process. Continuous processes such as those in the chemical industry use automatic controllers to adjust process factors in real time. Box and Luceño (1997) combine statistical process control with feedback process control techniques such as proportional/integral (PI) control. This reference (p. 133) adds that full adjustment corresponds to what Deming calls tampering, but use of a damping factor may avoid overadjustment.

The next step is to connect the normal distribution to the targets by placing the targets side by side with their corresponding bell curves. This is a good way to explain the concept to people without extensive mathematical backgrounds, and the control charts that correspond to the firearm targets will be added later.

The Normal Distribution

As stated previously, the normal distribution's probability density function (pdf) is

$$f(x) = \frac{1}{\sqrt{2\pi}\sigma} \exp\left(-\frac{1}{2\sigma^2}(x-\mu)^2\right)$$

where μ is the process mean and σ^2 is the variance. The process mean, μ, is therefore the point of aim while σ^2 is the spread in the shot pattern. Shorthand for this distribution is $N(\mu,\sigma^2)$, or normal with a mean of μ and a variance of σ^2. The distribution is symmetrical, so its mean (center of gravity), median (50th percentile), and mode (most likely value) are equal. Figure 1.3 shows the normal distributions that correspond to each shot pattern, where lower specification limit (LSL) and upper specification limit (USL) are the lower and upper specification limits, respectively.

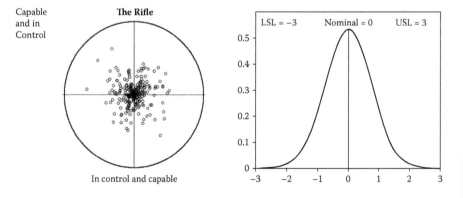

FIGURE 1.3
Normal distributions, capability, and control.

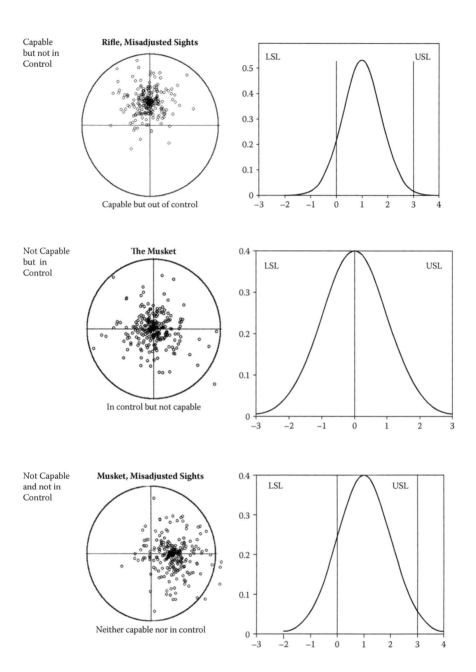

FIGURE 1.3
(Continued)

This section has illustrated the connection between a normal distribution's mean and variance with production yield. The next step is to understand *statistical hypothesis testing*. Control charts test the hypothesis that the process is in control, and there are always risks associated with such tests.

Statistical Hypothesis Testing

Hypothesis testing distinguishes genuine effects from random variation or luck. It is the foundation of design of experiments (DOE) as well as statistical process control.

All statistical experiments begin with the assumption of the *null hypothesis* (H_0), a starting assumption that is similar to the presumption of innocence in a criminal trial. One cannot prove the null hypothesis, just as acquittal in a trial is not proof of innocence. It means only that there is not enough evidence to accept the *alternate hypothesis* (H_1). Table 1.1 shows examples of the null and alternate hypothesis.

The presumption of innocence principle helps explain the concept of *significance level*. Statistical tests are often performed at a five percent significance level, and this can create some misleading ideas. Is there only a five percent chance that the experiment's results are significant?

Conviction of a defendant in a criminal trial requires him to be guilty beyond a reasonable doubt. Applied statisticians similarly reject the null hypothesis at a certain significance level or *quantified reasonable doubt*. Five percent significance level means that, if the null hypothesis is true (the defendant is innocent), there is only a five percent chance of its rejection.

The significance level is also known as the Type I risk, false alarm risk, alpha (α) risk, or producer's risk. The latter term is from acceptance sampling, and it is the producer's risk of rejecting a lot whose quality is acceptable. The tie-in with the risk of wrongly convicting an innocent defendant now becomes obvious.

TABLE 1.1

Null and Alternate Hypothesis

Null Hypothesis (H_0)	Alternate Hypothesis (H_1)
An experimental treatment has no effect; it is no different from the control. (Application: DOE)	The experimental treatment has an effect.
The manufacturing process is in control. (This application is this book's focus.)	The manufacturing process is out of control, and it needs adjustment.
A production lot meets the customer's quality requirements. (Application: acceptance sampling)	The production lot does not meet the customer's quality requirements.
The defendant is innocent.	The defendant is guilty.

The null hypothesis for statistical process control is the assumption that the process mean is at the nominal, and that the process variation has not changed from its historical level. The alternative hypothesis is that the mean has shifted or the variation has changed; in the latter case, the change is generally an undesirable increase. The significance level for the Shewhart control chart for the process mean is 0.00135 at each control limit, or 0.0027 for both control limits. The factory expects 2.7 false alarms per 1000 samples.

There is a corresponding Type II risk, beta (β) risk, or consumer's risk. The latter is also from acceptance sampling, and it is the risk of sending the customer a lot whose quality is unacceptable. It corresponds to the risk of acquitting a guilty defendant, and it is the risk of not rejecting H_0 when it should be rejected. The only way to reduce this risk without a corresponding increase in the false alarm risk is to increase the sample size.

A statistical test's *power*, $\gamma = 1 - \beta$, is its ability to reject H_0 when H_0 should be rejected. In statistical process control, this is its ability to detect a given change in the process mean or standard deviation. Table 1.2 summarizes statistical testing risks.

TABLE 1.2

Statistical Testing Risks: Decision vs. State of Nature

State of Nature → Decision	H_0 Is True	H_1 Is True and H_0 Should Be Rejected
Accept H_0 Acceptance of the null hypothesis really means we can't reject it because we don't have enough evidence to the contrary. Statistical tests never prove H_0.	Chance = $1 - \alpha$ • Acquit innocent defendant • Accept good lot (at the acceptable quality level in acceptance sampling) • Conclude there is no difference between experiment and control • Conclude that the process is in control in SPC • Boy doesn't cry "Wolf!"	Chance = β • Acquit guilty defendant • Accept bad lot (at the rejectable quality level or lot tolerance percent defective in acceptance sampling) • Miss a genuine difference between the experimental treatment and the control • Miss an out-of-control situation (assignable or special cause) in SPC • Boy doesn't see the wolf
Reject H_0	Chance = α (significance level, Type I risk, consumer's risk, false alarm risk) • Convict innocent defendant • Reject good lot • Conclude wrongly that the experimental treatment differs from the control • Conclude wrongly that the process is out of control • Boy cries "Wolf!"	Chance = $1 - \beta = \gamma$ (power) • Convict guilty defendant • Reject bad lot • Detect a genuine difference between the experimental treatment and the control • Detect an out-of-control process situation • Boy sees the wolf

The chance that the boy will see the wolf depends on how close the wolf gets to the sheep. The power of a designed experiment depends similarly on the difference between the control and experimental populations, and the power of a control chart depends on the deviation of a process's parameters from nominal. The power of an acceptance sampling test depends on how bad a production lot actually is, and the consumer's risk is specified for a specific level of badness: the rejectable quality level (RQL) or lot tolerance percent defective (LTPD). A control chart's *average run length* (ARL) is similarly the expected number of samples necessary to detect a given process shift, and it is the reciprocal of the power. If, for example, the power is 5 percent, the ARL is 20. A subsequent section in this chapter will show how to calculate it.

The next section will cover the traditional Shewhart control charts for normally distributed data. It will explain control limits and Western Electric zone tests in the context of false alarm (Type I) testing risks. These concepts will apply later to the development of charts for nonnormal distributions.

Control Chart Concepts

The purpose of control charts is to detect undesirable changes in the manufacturing process. These include changes in (1) variation and (2) accuracy.

1. Charts for variation: "Has a musket replaced the rifle?"
2. Charts for process mean: "Have the rifle's sights gone out of adjustment?"

The x-bar/R (sample average and sample range) and x-bar/s (sample average and sample standard deviation) charts test for changes in accuracy and variation. They require samples of at least two measurements. If only one measurement is available, either the X (individual measurements) chart or cumulative sum (CUSUM) chart must be used.

One might ask, "Why not take more than one measurement in all situations?" This leads to the concept of the *rational subgroup.*

Rational Subgroup

The rational subgroup was mentioned in the introduction to batch processes with nested variation sources. A rational subgroup represents a homogenous set of process conditions. Improper selection of the rational subgroup causes untold misery in factories that use statistical process control (SPC).

Consider a process that makes 55-gallon drum batches of a chemical. Specimens are taken from each drum for laboratory analysis, and the data are used for statistical process control purposes. The contents of each drum

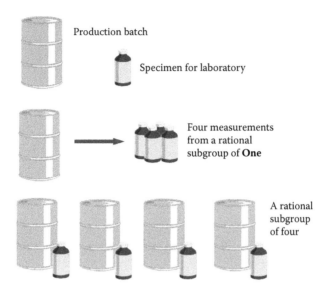

FIGURE 1.4
Rational subgroup for chemical batches.

should be homogenous (if the contents do not separate or settle), so within-batch variation should be small. There may, however, be considerable between-batch variation, so the product's composition may vary from drum to drum.

It is easy to make the mistake of calling four specimens from one drum a subgroup of four, as shown in Figure 1.4. They are no such thing; their standard deviation or range reflects only the within-drum variation, which is likely to be the least important variance component. Use of this information to calculate control limits will lead back to the situation shown in the introduction, where points violate the lower and upper control limits of the x-bar chart. The reason is that the sigma in the "mean +/–3 sigma" control limits reflects only the within-batch variance component, but the variation in the so-called sample average includes the larger between-batch component.

If the drums are tested one at a time, the X chart for individuals is appropriate for this process. This, incidentally, makes it impossible to rely on the central limit theorem if the quality characteristic (like an impurity level) does not follow the normal distribution.

Central Limit Theorem

The central limit theorem was mentioned previously, and this is a good place to discuss its exact meaning. It says that *the averages of infinite samples follow the normal distribution even if the underlying population is nonnormal.*

In practice, averages of four to ten measurements often behave sufficiently normally to allow problem-free use of the traditional x-bar Shewhart chart

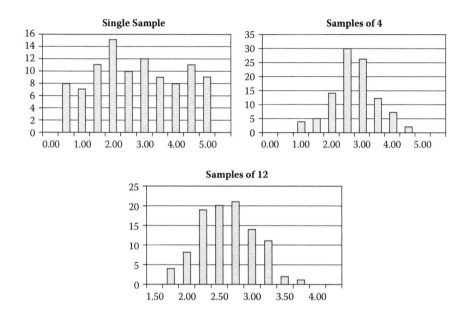

FIGURE 1.5
Central limit theorem.

for process mean. Figure 1.5 shows histograms for (a) 100 measurements from a uniform distribution on the interval [0,5], (b) 100 averages of four measurements from the same distribution, and (c) 100 averages of 12 measurements from the same distribution. The histogram for the averages of four begins to resemble a bell curve, and that for the averages of 12 is even more bell shaped. (Note the change in axis scale for the latter histogram.)

As shown in the chemical batch example, however, averages might not be available. Furthermore, *specifications refer to individual measurements and not sample averages.* Customers won't accept nonconforming work because on average, the five units in the sample are in specification.

The rational subgroup concept supports understanding of the traditional Shewhart control charts. The next step is to set up the charts and use them.

Setup and Deployment of Control Charts

The first step is to collect data from the process and use it to estimate the mean and standard deviation. This is sometimes referred to as Phase I, in which a historical dataset (HDS) is selected and possibly purged of outliers before estimation of the process parameters and control limits. Phase II consists of routine use of the resulting control limits on the shop floor.

The data are subject to the following requirements:

1. The rational subgroup must be defined properly.
2. The process must be in control, and this can be determined from the control chart itself.
   - Since the null hypothesis that the process is in control can never be proven, it may be more accurate to say that the control chart does not show beyond a reasonable doubt that the process is not in control.
3. The data must follow whatever distribution is selected to model the process. This chapter describes tests for goodness of fit that *must* be performed as part of any process capability analysis or control chart setup.

The sample average, range, and standard deviation are defined as follows. For the *i*th subgroup of n_i measurements,

Equation Set 1.1

$$\bar{x}_i = \frac{1}{n_i} \sum_{j=1}^{n_i} x_{i,j} \quad R_i = \max(x_i) - \min(x_i) \quad s_i = \sqrt{\frac{1}{n_i - 1} \sum_{i=1}^{n_i} (x_{i,j} - \bar{x}_i)^2}$$

There are two ways to define the process mean and standard deviation. *Theoretical control charts* assume that the process population's mean and variance are known; it is given that the process population follows the distribution $N(\mu, \sigma^2)$. This could be a valid assumption for processes with which a factory has had long experience.

Empirical control charts require estimation of the process parameters from data. The process population follows the distribution $N(\hat{\mu}, \hat{\sigma}^2)$ where the carat or "hat" (^) means *estimated parameter*. Theoretical and empirical charts work similarly, but the calculation of their control limits uses different control chart factors (Table 1.3). These are tabulated (Appendix A) functions of the sample size (*n*). When the sample size is not consistent, American Society for Testing and Materials (ASTM, 1990, 69) estimates the process standard deviation from the sample standard deviations with the expression

$$\hat{\sigma} = \frac{1}{m} \sum_{i=1}^{m} \frac{s_i}{c_4(n_i)}$$

where n_i is the size of the *i*th sample, *m* is the total number of subgroups, and $c_4(n_i)$ is the c_4 factor for the indicated sample size. The estimate from the sample ranges is similar (ASTM, 1990, 70).

Consider the data in Table 1.4, which is a simulation of a normal distribution with mean 50 and variance 4 (Control_Chart_Fixed_Sample in the Excel

TABLE 1.3

Control Chart Formulas

	Theoretical Charts: σ Is Given	Empirical Charts: Estimate σ from Data
x-bar/R charts:	Given	For m samples of size n_i,
Estimate for process standard deviation σ		$\hat{\sigma} = \frac{1}{m} \sum_{i=1}^{m} \frac{R_i}{d_2(n_i)}$
Control limits, R chart for process variation	$[D_1\sigma, D_2\sigma]$ Center line: $d_2\sigma$ (see note)	$[D_3\bar{R}, D_4\bar{R}]$ Center line: $\bar{R}$
Control limits, x-bar chart for process mean	$\mu \pm 3\dfrac{\sigma}{\sqrt{n}}$	$\bar{\bar{x}} \pm A_2\bar{R}$
x-bar/s charts:	Given	For m samples of size n_i,
Estimate for process standard deviation σ		$\hat{\sigma} = \frac{1}{m} \sum_{i=1}^{m} \frac{s_i}{c_4(n_i)}$
Control limits, s chart for process variation	$[B_5\sigma, B_6\sigma]$ Center line: $c_4\sigma$ (see note)	$[B_3\bar{s}, B_4\bar{s}]$ Center line: $\bar{s}$
Control limits, x-bar chart for process mean	$\mu \pm 3\dfrac{\sigma}{\sqrt{n}}$	$\bar{\bar{x}} \pm A_3\bar{s}$

Note: The B, c, D, and d factors are all functions of sample size. When the sample size varies, the center line for the R or s chart will actually move.

file of simulated data). The following Excel spreadsheet functions calculate the required values for row 6.

$$\text{x\_bar} = \text{AVERAGE(B6:E6)}$$

$$R = \text{MAX(B6:E6)} - \text{MIN(B6:E6)}$$

$$s = \text{STDEV(B6:E6)}$$

Some caution is required for use of these functions on large samples, *because they work for a maximum of 30 numbers*. STDEV will not, for example, return the standard deviation of the entire set of 100 numbers, nor will the MIN and MAX functions return the smallest and largest members of the dataset. The same applies to the COUNT function, so it is not reliable for counting data when there are more than 30 samples.

Then estimate the mean and standard deviation as follows:

$$\hat{\mu} = 50.012$$

$$\hat{\sigma} = \frac{\bar{R}}{d_2(4)} = \frac{4.085}{2.059} = 1.984$$

$$\hat{\sigma} = \frac{\bar{s}}{c_4(4)} = \frac{1.824}{0.9213} = 1.980$$

TABLE 1.4

Normally Distributed Data $X \sim N(50,4)$

Sample	X1	X2	X3	X4	x_bar	R	s
1	46.768	47.762	50.949	50.896	49.094	4.180	2.150
2	48.765	51.639	51.300	49.711	50.354	2.874	1.352
3	52.052	48.185	51.885	50.291	50.603	3.868	1.797
4	46.387	47.312	49.872	47.974	47.887	3.485	1.475
5	50.879	51.393	48.311	49.911	50.124	3.081	1.355
6	47.864	47.598	49.449	49.066	48.494	1.851	0.902
7	51.321	52.265	50.935	46.118	50.160	6.147	2.752
8	48.181	47.444	51.604	47.256	48.621	4.348	2.028
9	50.122	49.961	49.587	50.457	50.032	0.871	0.362
10	49.526	52.702	48.206	50.268	50.175	4.495	1.888
11	51.137	49.667	47.904	49.734	49.611	3.234	1.325
12	50.420	50.018	53.317	52.527	51.570	3.299	1.602
13	53.232	49.372	47.718	52.746	50.767	5.514	2.660
14	49.811	50.025	50.328	50.479	50.161	0.668	0.300
15	46.765	45.497	51.101	50.832	48.549	5.604	2.841
16	52.304	52.880	49.942	51.650	51.694	2.938	1.272
17	54.162	49.018	50.148	47.707	50.259	6.455	2.787
18	49.791	54.295	50.891	51.726	51.676	4.505	1.918
19	49.122	52.835	49.087	47.099	49.536	5.736	2.394
20	56.413	47.080	50.964	51.079	51.384	9.334	3.834
21	50.044	50.957	49.339	48.769	49.777	2.188	0.943
22	49.550	51.908	49.839	47.215	49.628	4.694	1.921
23	50.100	48.819	50.275	48.824	49.504	1.456	0.792
24	49.111	49.266	51.117	52.535	50.507	3.424	1.630
25	49.354	54.780	46.912	49.459	50.127	7.868	3.318
			Grand Average		50.012	4.085	1.824

The control limits based on range are

$$\bar{\bar{x}} \pm A_2 \bar{R} = 50.012 \pm 0.729 \times 4.085 = [47.034,\ 52.990]$$

$$[D_3 \bar{R}, D_4 \bar{R}] = [0,\ 2.282 \times 4.085] = [0,\ 9.322]$$

Figure 1.6 shows a Minitab-generated x-bar/R chart for 100 subgroups of four measurements from the distribution $N(50,2)$. Control limits based on standard deviation are

$$\bar{\bar{x}} \pm A_3 \bar{s} = 50.012 \pm 1.628 \times 1.824 = [47.043,\ 52.981]$$

$$[B_3 \bar{s}, B_4 \bar{s}] = [0, 2.266 \times 1.824] = [0,\ 4.133]$$

Figure 1.7 shows an x-bar/s chart for the same data.

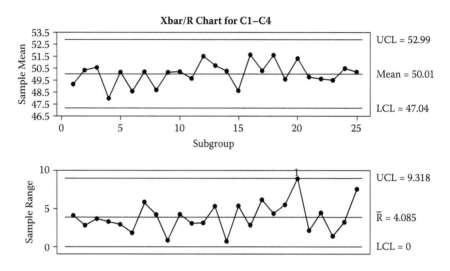

FIGURE 1.6
x-bar/R chart.

x-bar/R Chart and x-bar/s Chart

The x-bar/R chart is actually a holdover from the days of manual calculations. The sample standard deviation (s) is slightly more powerful than the sample range (R) in terms of its ability to detect a change in process variation, but it was too computation intensive for practical shop floor use in the days before computers. The range is simply the difference between the largest and smallest measurements

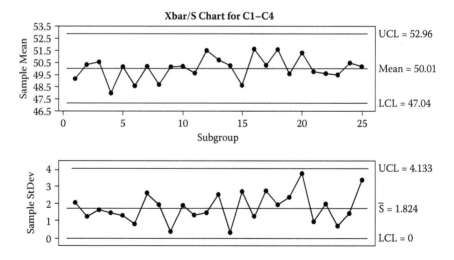

FIGURE 1.7
x-bar/s chart.

in the sample, and it is suitable for easy hand calculation. Figures 1.6 and 1.7 are x-bar/R and x-bar/s charts as prepared by Minitab 13.

Matters become somewhat more complicated when the sample sizes are not equal. Our experience is that many commercial SPC packages cannot handle these situations, but Minitab 15 and StatGraphics Centurion can. For the latter, deselect the check box **Use average subgroup sizes** under the **Analysis Options.**

Treatment of Variable Subgroup Sizes

An Excel spreadsheet can handle the necessary calculations for variable subgroup sizes as shown in Control_Chart_Variable_Sample in the simulated data spreadsheet. In addition, conditional cell formatting can turn the cells for the sample average, range, and/or standard deviation red if they are outside their control limits. This *visual control* makes the status of the process obvious to the production worker. Note also that the spreadsheet calculates control limits from the standards for the mean and standard deviation as opposed to the current grand average and estimates of the standard deviation. No control chart should ever change its control limits in response to additional measurements.

Sample calculations are as follows for row 10; results may not match those of the spreadsheet exactly due to manual calculation with the indicated number of figures. Given a sample average of 48.493, range of 4.1802, and standard deviation of 2.1839,

$$\hat{\sigma}_R = \frac{4.1802}{1.693} = 2.469$$

$$\hat{\sigma}_s = \frac{2.1839}{0.8862} = 2.464$$

If 1.969 is the (empirical) standard deviation, then the control chart center lines are as follows for a sample of 3:

$$\bar{R} = \hat{\sigma} \times d_2(3) = 1.969 \times 1.693 = 3.334$$

$$\bar{s} = \hat{\sigma} \times c_4(3) = 1.969 \times 0.8862 = 1.745$$

The control limits are

$$R : [D_3 \bar{R}, D_4 \bar{R}] = [0,\ 2.575 \times 3.334] = [0,\ 8.585]$$

$$s : [B_3 \bar{s}, B_4 \bar{s}] = [0,\ 2.568 \times 1.745] = [0,\ 4.481]$$

Figure 1.8 shows how Minitab 15 handles the dataset in question, and Figure 1.9 shows the results from StatGraphics Centurion. The most important feature consists of control limits that depend on the sample size, and a center line for the R or s chart that varies with sample size.

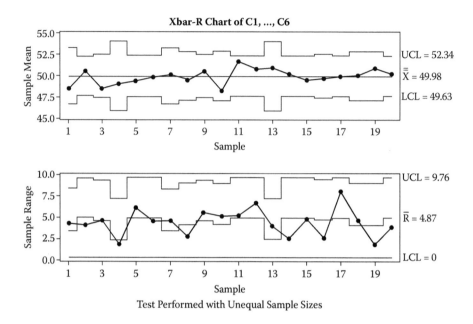

FIGURE 1.8
Minitab x-bar/R chart for variable sample sizes.

Minitab's estimate of the process standard deviation based on sample ranges is 1.922; it is important to select **Rbar** versus **Pooled Standard Deviation** to get this result. StatGraphics Centurion also gets 1.922.

StatGraphics and Minitab both get 1.905 for the estimate of the process standard deviation from the sample standard deviations. These programs' empirical estimates for the process standard deviation differ slightly from those from the spreadsheet, and this discrepancy was investigated by testing the example from the ASTM reference (1990, Table 30, 70). The spreadsheet reproduces the estimate of the standard deviation based on range, 0.812, exactly. If the sample standard deviations are rounded to the nearest hundredth as they are in the reference, the spreadsheet returns the same estimate of standard deviation based on sample standard deviations, 0.902. In other words, the spreadsheet reproduces the ASTM example's results correctly.

In contrast, Minitab and StatGraphics Centurion both return 0.802. The documentation from StatGraphics shows that it uses a different method to estimate the process standard deviation. For k subgroups and from the sample ranges, StatGraphics uses

$$\hat{\sigma} = \frac{\sum_{j=1}^{k} \frac{f_j R_j}{d_2(n_j)}}{\sum_{j=1}^{k} f_j} \quad \text{where } f_j = \frac{d_2^2(n_j)}{d_3^2(n_j)}$$

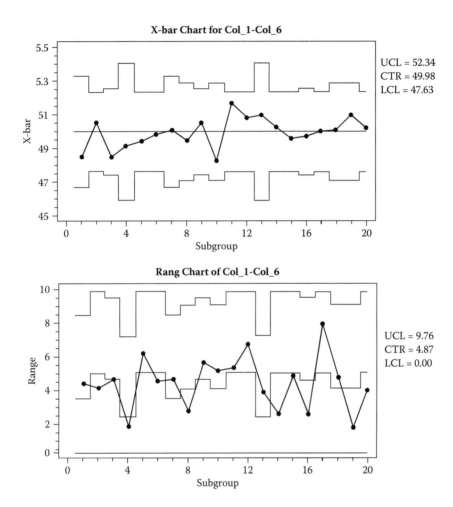

FIGURE 1.9
Statgraphics x-bar/R chart for variable sample sizes.

From sample standard deviations with bias correction:

$$\hat{\sigma} = \frac{\sum_{j=1}^{k} \frac{h_j s_j}{c_4(n_j)}}{\sum_{j=1}^{k} h_j} \quad \text{where} \quad h_j = \frac{c_4^2(n_j)}{1 - c_4^2(n_j)}$$

Although the traditional control limits for the sample range and standard deviation charts are mean ±3 sigma limits, the one-sided false alarm risk is not the expected 0.135 percent because these sample statistics do not follow the normal distribution. Chapter 3 shows how to calculate R chart control limits with known and exact false alarm risks, and the 0.00135 and 0.99865 quantiles are given for this statistic. This chapter will later treat the chi square chart for sample standard deviations, and the control limits of this chart can easily be set to provide known and exact false alarm risks. No matter what method is used to set up the chart, however, it is necessary to interpret and understand the chart. That is the subject of the next section.

Interpretation of x-bar/R and x-bar/s Charts

The next step is to interpret the charts, and shot patterns on targets are useful for explaining variation and accuracy. Figure 1.10 places x-bar/s charts side by side with targets and histograms to show how each chart should be interpreted.

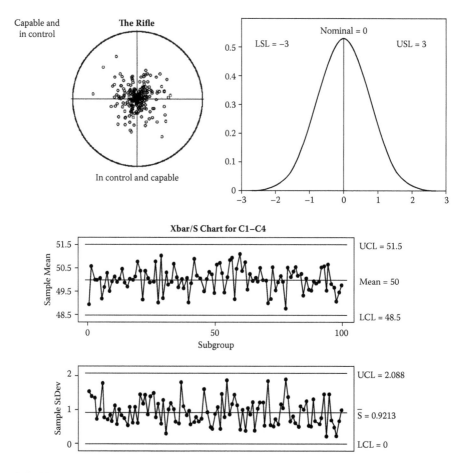

FIGURE 1.10
Control chart interpretation.

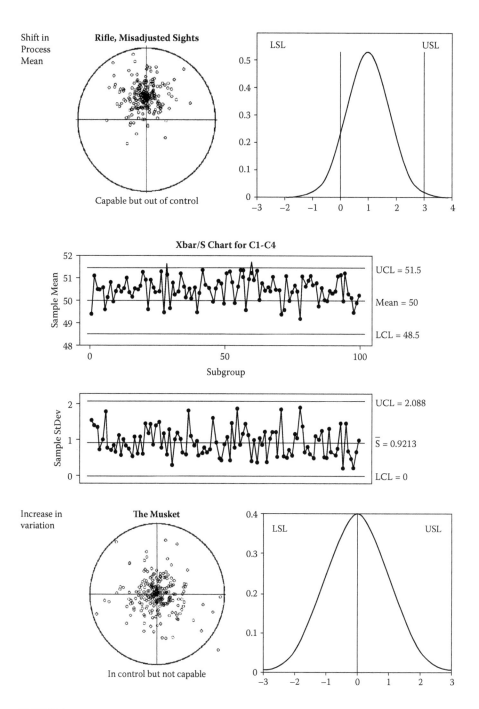

FIGURE 1.10
(Continued)

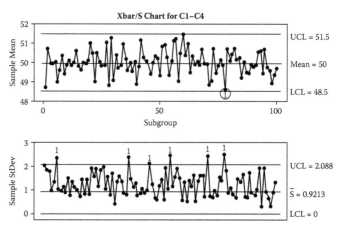

The x-bar (sample #78) outside the control limits is almost certainly due to the increase in variation. The chart's control limits assume σ=1.00, but σ=1.33 and there is more spread in the sample averages. Since the control limits for the x-bar chart depend on assumptions about the standard deviation, the variation chart is interpreted first in situations like this one. When points are outside the control limits for both charts, the assignable cause is assumed to involve an incrase in variation as opposed to a shift in process mean.

Change in mean and increse in variation

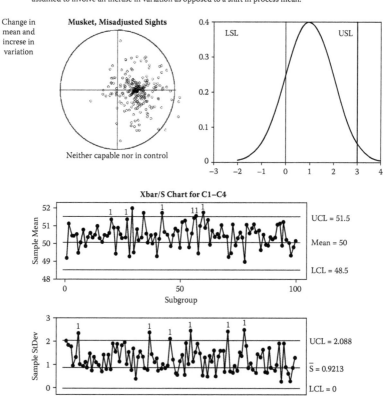

FIGURE 1.10
(Continued)

A *decrease* in variation is a process improvement, and is very unlikely to occur unless someone does something to make it happen.

Treatment of Out-of-Control Signals

A control chart is considered out of control when the first point exceeds a control limit. It is a mistake to continue to run the process and see whether the chart comes back into control. Most points will be within the control limits even when the process's mean or variation has changed! The Type II (consumers, beta) risk inherent in any statistical test is the chance that the boy won't see the wolf, and in statistical testing, this risk is very substantial unless the wolf is very close to the sheep. In manufacturing terms, the process will have to be very bad before even half the measurements violate control limits; a condition under which it will almost certainly make a lot of scrap. The purpose of SPC is to allow correction of the process before it makes any scrap at all.

All out-of-control signals require investigation. It is possible that the investigation will reveal nothing because there is the Type I (producer's, alpha) risk of a false alarm. The x-bar chart will in fact have an average of 2.7 false alarms for every thousand samples. If the investigation finds no assignable cause for the out-of-control signal, this must still be recorded.

An *out-of-control action procedure* (OCAP) or corrective and preventive action matrix (CP matrix per Hradesky, 1988) should be part of the work instruction. It tells the operator what to do when a point exceeds the control limits. There is often a checklist of steps the operator can take to bring the process back into control without having to call an engineer or technician, and this is consistent with standards for quality management systems.

SPC and ISO 9001:2008

This is a good place to discuss the relationship of SPC and the International Organization for Standardization (ISO) 9001:2008 standard for quality management systems. ISO 9000 is best described as "say what you do and do what you say." "Say what you do" means having a quality management system (QMS) with documented procedures for performing jobs, and this QMS must fulfill the requirements for the ISO 9000 standard. "Do what you say" means following the documented procedures that your organization has created. Failure to act on an out-of-control signal is an ISO 9000 nonconformance (not doing what you say). If an auditor sees a control chart with such a signal and no evidence of investigation or corrective action, he or she should record a nonconformance.

A control chart for a process that makes customer-shippable products is a *quality record*. This is a fourth-tier document[2] that *must* be subordinate to a third-tier work instruction or operating instruction. Only a work instruction can tell anyone how to do a job, so a work instruction must cite the form or control chart in question (and vice versa). In this case the work instruction for the job must call for the control chart to be kept as part of the job's process control activity. Furthermore, as a quality record, the control chart must be protected from loss (e.g., a computer hard drive crash) according to the quality system's record retention procedures.

It is acceptable to have control charts (or trend charts) whose only purpose is to gather information. They are not auditable under ISO 9000 because they have no role in controlling the process or assuring the quality of the product. Such charts should be clearly labeled "for information only," "not for process control," or something to that effect.

X (Individual Measurement) Charts

As shown in Figure 1.4, it is sometimes not possible to get a rational subgroup of more than one measurement. To set the control limits for the X chart, estimate the standard deviation from the average moving range as follows.

Equation Set 1.2

$$\overline{MR} = \frac{1}{n-1}\sum_{i=2}^{n}|X_i - X_{i-1}| \quad \hat{\sigma} = \frac{\sqrt{\pi}}{2}\overline{MR} \Rightarrow \text{control limits } \overline{\overline{x}} \pm 2.66\overline{MR}$$

The moving range (MR) chart has no real value, and it need not be plotted. Messina (1987, 132) says that the MR chart contains no information that is not present in the X chart. ASTM (1990, 97) says, "All the information in the chart for moving ranges is contained, somewhat less explicitly, in the chart for individuals." AT&T[3] (1985, 22) says, "Do not plot the moving ranges calculated in step (2)."

Figure 1.11 shows an X chart as prepared by Minitab. "I" stands for individuals. One hundred measurements have an average of 49.956 and an average moving range of 1.267. The estimate for sigma is therefore 1.123 and the control limits are [46.59, 53.33].

Recall that any statistical hypothesis test has two risks: that of a false alarm (the boy who cries "Wolf!") and the risk of failure to detect an undesirable condition (the boy does not see a genuine wolf). The power of the chart for individuals is less than an x-bar chart for samples of two or more, and Montgomery (1991, 280) recommends the CUSUM chart as a particularly effective alternative. The power of the chart for process averages increases with sample size, as the next section on average run length will demonstrate.

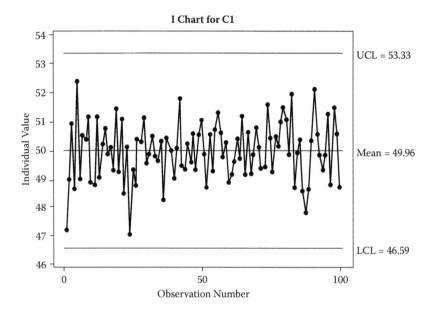

FIGURE 1.11
X chart for individual measurements.

Average Run Length (ARL)

Given a process shift of δ standard deviations in a process that follows the normal distribution, the chance that the average of n pieces will be outside the nearest control limit is as follows (with the upper control limit [UCL] as an example). μ_0 is the hypothetical or nominal process mean.

$$\gamma(\delta, n) = 1 - \Pr(\bar{x} \leq UCL) = 1 - \Phi\left(\frac{UCL - (\mu_0 + \delta\sigma)}{\frac{\sigma}{\sqrt{n}}}\right)$$

This is simply one minus the quantile of the upper control limit given the new process mean, and the ARL is the reciprocal of the resulting power γ. For an x-bar chart,

$$\gamma(\delta, n) = 1 - \Phi\left(\frac{\left(\mu_0 + 3\frac{\sigma}{\sqrt{n}}\right) - (\mu_0 + \delta\sigma)}{\frac{\sigma}{\sqrt{n}}}\right) = 1 - \Phi(3 - \delta\sqrt{n}) \qquad (1.3)$$

Figure 1.12 illustrates the concept graphically for a process whose nominal mean and standard deviation are 50 and 2, respectively. A sample of four pieces will therefore follow the distribution $N(50,1)$ if the process is in control,

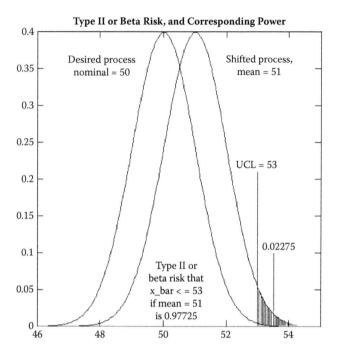

FIGURE 1.12
Type II risk and power of a Shewhart control chart.

and the control limits of the x-bar chart are 47 and 53. If the process mean shifts by 1, there is a

$$\Phi\left(\frac{53-51}{\frac{2}{\sqrt{4}}}\right) = 0.97725$$

chance that the resulting sample mean will be below the upper control limit of 53. This is the Type II risk or β risk. The corresponding chance that the sample average will be outside the control limits is 0.02275, and the ARL is its reciprocal. It will require an average of 44.0 samples to detect the process change in question.

Figure 1.13 shows the average run lengths as a function of the mean's shift in standard deviations for samples of 1, 4, and 9. The 1.5 sigma shift that Six Sigma assumes in the calculation of 3.4 defects per million opportunities would be detected, on average, by two samples of four. The ARL is meanwhile 740.80 for no process shift, and this corresponds to the false alarm risk of 0.00135 for each control limit.

This approach can be generalized for other distributions in which the sample statistic is an average, range, or standard deviation. The beta or Type II risk at the upper control limit (UCL) is always the quantile of the UCL given the changed distribution parameters, and that at the lower control limit (LCL) is 1 minus the quantile of the LCL.

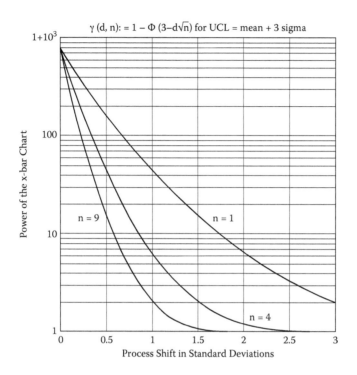

$\gamma (d, n): = 1 - \Phi (3-d\sqrt{n})$ for UCL = mean + 3 sigma

Power of the x-bar Chart

$n = 9$ $n = 1$

$n = 4$

Process Shift in Standard Deviations

FIGURE 1.13
Average run length, x-bar chart.

z Chart for Sample Standard Normal Deviates

Suppose a machine tool makes different parts, each of which has a different nominal measurement (and possibly historical standard deviation). This supports mixed-model production, which is an important objective in many lean manufacturing initiatives. The idea is to make parts in the order, for example, A-A-B-C-A-A-B-C as opposed to long runs of A followed by shorter runs of B and C. This prevents accumulation of inventory because only small quantities of each part are made at one time. It is not practical to keep a control chart for each part, but the z chart allows a single chart to track all the products from a given piece of equipment.

Standard normal deviate is a lot less intimidating in plain English: the number of standard deviations between the number of interest and the mean. For mixed-model production in which part k has a nominal (or historical mean) of μ_k and standard deviation σ_k,

$$z = \frac{x - \mu_k}{\sigma_k} \text{ or, more generally, } z = \frac{\bar{x} - \mu_k}{\frac{\sigma_k}{\sqrt{n}}} \tag{1.4}$$

$z \sim N(0,1)$; the standard normal deviate has by definition a mean of zero and a variance of one. The resulting z values can therefore be tracked on a single control chart whose center line is zero and whose control limits are $[-3,3]$.

The χ^2 Chart (or s^2 Chart) for Sample Standard Deviations

The χ^2 chart does for the sample standard deviation what the z chart does for the sample average. Let

$$\chi^2 = \frac{(n-1)s^2}{\sigma_{0,k}^2}$$

where s^2 is the sample's variance and $\sigma_{0,k}^2$ is the nominal or expected variance for product k. The control limits for χ^2 are the $(\alpha/2)$ and $(1 - \alpha/2)$ quantiles of the chi square distribution with $n - 1$ degrees of freedom. α is the desired total false alarm risk, which is 0.00270 for traditional Shewhart charts. The chart's center line is the median (0.50 quantile) of this chi square distribution. Note, incidentally, that these limits are more accurate than the traditional ones for the s chart because the latter only approximate the 0.00135 and 0.99865 quantiles. The user can in fact select any desired false alarm risk for the chi square chart. If tabulated values are not available for the desired quantile, Excel's CHIINV function can return it.

Table 1.5 shows just how easy it is to automate the entire process on a spreadsheet and to accommodate variable sample sizes as well. It is relatively easy to add a *visual control*, a signal that makes the status of the process obvious, that turns the cell for z or χ^2 red if the result is outside the control limits. The mixed-model production sequence A-A-B-C is used.

Table of product nominal means and sigmas:

	R	S	T
1		Nominal	Nominal
2	Product	Mean	Sigma
3	A	50	1
4	B	75	1.5
5	C	100	2

Calculations (copy to subsequent rows):

G4 (n) = COUNT(B4:F4)

H4 (μ_0) = VLOOKUP($A4,$R$3:$T$5,2)

I4 (σ_0) =VLOOKUP($A4,$R$3:$T$5,3)

J4 (x-bar) = AVERAGE(B4:F4)

K4 (s) = STDEV(B4:F4)

L4 (z) = (J4−H4)/(I4/SQRT(G4))

M4 (χ^2) = (G4−1)*K4^2/I4^2

N4 (χ^2 LCL) = CHIINV(0.99865,$G4−1)

O4 (χ^2 CL) = CHIINV(0.5,$G4−1)

P4 (χ^2 UCL) = CHIINV(0.00135,$G4−1)

Sample calculations, row 5 (second sample): x-bar = 50.249, s = 0.9073. Note that Excel's AVERAGE and STDEV functions automatically detect the number

TABLE 1.5

z and χ^2 Charts on an Excel Spreadsheet

	A	B	C	D	E	F	G	H	I	J	K	L	M	N	O	P
2								Nominal	Nominal							
3	Product	X1	X2	X3	X4	X5	n	Mean	Sigma	x-bar	s	z	Chi sq.	LCL	CL	UCL
4	A	47.212	48.988	50.937	48.657	52.404	5	50	1	49.640	2.037	−0.806	16.604	0.106	3.357	17.800
5	A	48.995	50.468	50.373	51.162		4	50	1	50.249	0.907	0.499	2.468	0.030	2.366	15.630
6	B	70.819	73.482	76.406	72.986		4	75	1.5	73.423	2.300	−2.102	7.056	0.030	2.366	15.630
7	C	94.425	97.976	101.875			3	100	2	98.092	3.726	−1.652	6.942	0.003	1.386	13.215
8	A	47.212	48.988				2	50	1	48.100	1.256	−2.687	1.577	0.000	0.455	10.273
9	A	48.995	50.468	50.373	51.162	48.824	5	50	1	49.964	1.012	−0.080	4.093	0.106	3.357	17.800
10	B	70.819	73.482	76.406	72.986		4	75	1.5	73.423	2.300	−2.102	7.056	0.030	2.366	15.630
11	C	94.425	97.976	101.875	97.314	104.807	5	100	2	99.279	4.075	−0.806	16.604	0.106	3.357	17.800

of measurements. The blank cell is simply ignored. The COUNT function determines that $n = 4$. The VLOOKUP function cross-references the product A against the table in R1:T5 to get the nominal mean (50) and standard deviation (1) for this product.

$$z = \frac{50.249 - 50}{\frac{1}{\sqrt{4}}} = 0.498 \quad \text{and} \quad \chi^2 = \frac{(4-1) \times 0.9073^2}{1^2} = 2.470$$

(These do not match the tabulated values exactly because the spreadsheet carries all the significant figures and the hand calculation uses only those shown.) The control limits for the χ^2 chart are the 0.00135 and 0.99865 quantiles of the χ^2 distribution with $4 - 1 = 3$ degrees of freedom, and the centerline is this distribution's median.

Acceptance Control Chart

The acceptance control chart (Feigenbaum, 1991, 426) is useful when it is not possible or convenient to hold the process mean at nominal. We encountered a practical application in which incremental changes to a process were not possible, so adjustment in response to an out-of-control signal could move the mean even farther from nominal. The effects would have been similar to what is known traditionally as overadjustment, even though the process technically needed adjustment. The process owner was satisfied if the mean remained within a certain range, and the acceptance control chart is ideal for applications of this nature. Nothing is free in applied statistics, and the price of the allowable "play" in the process mean consists of lower process capability. Another way to say this, however, is that lower process capability is the unavoidable consequence of inability to adjust the process mean in small increments.

Let the acceptable range for the mean be the lower and upper *acceptable process levels* (LAPL,UAPL). These limits define a center band, as opposed to a center line, for the control chart. The acceptance control limits are then

$$LAPL - \frac{z_\alpha \sigma}{\sqrt{n}}, \quad UAPL + \frac{z_\alpha \sigma}{\sqrt{n}}$$

where α is the desired false alarm risk; for $\alpha = 0.00135$, $z = 3$ which yields traditional Shewhart limits.

The reference defines the acceptable process levels as $LAPL = LTL + z_{p1}\sigma$, $UAPL = UTL - z_{p1}\sigma$ where $p1$ is the expected nonconforming fraction when the process operates at either limit of [LAPL,UAPL]. A similar set of calculations defines rejectable process levels, and the space between the rejectable and acceptable process levels is the *indifference zone*. The latter are not necessary for the chart the operator sees, and it is even unnecessary to include the specification limits that appear in Figure 1.14.

Note again that the process capability or process performance indices will be lower than those of a process that operates at the nominal. The lower

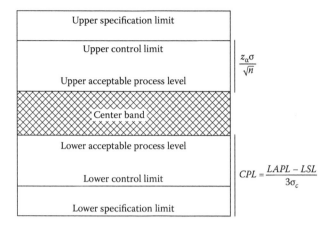

FIGURE 1.14
Acceptance control chart.

capability index (CPL) is, for example,

$$CPL = \frac{LAPL - LSL}{\sigma_c} < \frac{\mu_0 - LSL}{\sigma_c}$$

where μ_0 is the nominal and σ_c is the standard deviation as estimated from subgroup statistics.

Western Electric Zone Tests

The Western Electric zone tests are additional tests whose purpose is to detect changes in the process mean. They can also be used (with considerably less accuracy in terms of Type I risks) on the charts for process variation, although Chapter 3 will show how to calculate boundaries with known and exact false alarm limits for the R and s charts.

Figure 1.15 shows the rules for the Western Electric zone tests. The Shewhart control chart is divided into three zones on each side of the center line. Zone C consists of everything to one side of the center line. Zone B is one or more standard deviations from the center line. Zone A is two or more standard deviations from the center line.

The following tests (Table 1.6) are optional, and each one increases the overall Type I (false alarm) risk. They can, however, detect an out-of-control situation more quickly than the use of the traditional 3-sigma control limits alone. The 3-sigma control limit criterion is not a zone test, but it is included for comparison.

The rationale behind the Zone C test is easiest to understand. If the process is in control—that is, the process mean equals the chart center line—there is a 50 percent chance of getting a point on either side of the centerline. Eight

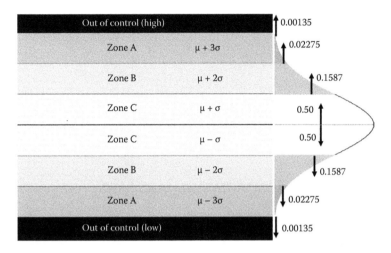

FIGURE 1.15
Western Electric zone tests.

consecutive points above the line are like throwing a coin and getting eight heads in a row. The chance of doing this is 0.5 to the eighth power, or 0.00391. This is also the chance of getting eight consecutive points below the center line, or throwing eight tails in a row. In the case of an actual coin, these results would suggest strongly that the coin was loaded. On a control chart, they suggest that the process mean is *not* on the chart centerline.

The cumulative binomial distribution must be used to calculate the Type I risk for Zones A and B.

$$\text{Zone A: } \frac{3!}{2!1!}0.0022752^2 \times (1-0.02275) + \frac{3!}{3!0!}0.0022753^3 = 0.00153$$

$$\text{Zone B: } \frac{5!}{5!1!}0.1587^4 \times (1-0.1587) + \frac{5!}{5!0!}0.1587^5 = 0.00277$$

TABLE 1.6

Western Electric Zone Tests

Test	Criteria: Points Must Be on the Same Side of the Center Line (+ or −)	Type I Risk (1-Sided Test)	Type I Risk (2-Sided Test)
Control limits	1 point outside a control limit	0.00135	0.00270
Zone A	2 (or more[a]) out of 3 consecutive points in Zone A	0.00153	0.00306
Zone B	4 (or more) out of 5 consecutive points in Zone B	0.00277	0.00554
Zone C	8 consecutive points on one side of the center line	0.00391	0.00781

[a] Used in calculating the false alarm risk. The idea is that there are either 2 or 3 out of the next three points in Zone A waiting to be found. In the latter case, the process would be called out of control when the second point was plotted. The same reasoning applies to Zone B.

It is important to remember that anything that requires somebody to read a vernier or, in this case, count points in each zone, is not likely to get done consistently—which in turn becomes an ISO 9001 nonconformance if the work instruction calls for it. These tests should probably not be used unless the software can alert the operator when a failing result occurs. In contrast, a point that is actually outside a control limit is unambiguous and obvious, and it takes next to no time to observe it. Table 1.7 offers a simple visual control that might be deployed on a spreadsheet.

TABLE 1.7

Zone Tests on a Spreadsheet

	A	B	C	D	E	F	G	H	I
2	Mean	100							
3	Sigma	2							
4									
5	X1	X2	X3	X4	X5	n	x_bar	Chart	z
6	97.48	99.67	97.24	104.68	102.16	5	100.25	|C	0.28
7	100.25	102.56	97.93			3	100.25	|C	0.21
8	100.19	98.58	100.39	97.45		4	99.15	C|	−0.85
9	99.58	100.55	101.99	100.13	99.83	5	100.42	|C	0.46
	101.47	100.73	101.09			3	101.09	|C	0.95
	104.00	99.36	103.19			3	102.18	|B	1.89
	98.46	97.66	97.06	99.64	101.32	5	98.83	B|	−1.31
	97.42	98.41	100.47	96.33		4	98.16	B|	−1.84
	102.39	99.61	97.54	96.25	97.93	5	98.74	B|	−1.40
	98.17	98.82	102.00	101.91	100.45	5	100.27	|C	0.30
	99.31	100.96				2	100.14	|C	0.10
	99.05	98.77	101.10	100.41	100.68	5	100.00	|C	0.00
	98.73	98.02	98.16	98.59	99.64	5	98.63	B|	−1.53
	97.38	102.09	99.91	102.64		4	100.50	|C	0.50
	98.39	100.95	101.22	100.48		4	100.26	|C	0.26

	K	L
	z	Text
5		
6	−5	X |
7	−4	X |
8	−3	A |
9	−2	B |
10	−1	C |
11	0	|C
12	1	| B
13	2	| A
14	3	| X
15	4	| X

Column I is actually not necessary, and it appears only to help the reader interpret column H. The entry in column H6 is = VLOOKUP (((G6–$B$2)* SQRT(F6)/$B$3,K$6:L$15,2) and it calculates

$$z = \frac{\bar{x} - \mu}{\frac{\sigma}{\sqrt{n}}}.$$

The resulting visual control shows the zone into which the sample statistic falls, and a strong argument can be made that no control chart at all is then necessary. It is in fact easier to see a zone test failure (such as four " B|" results for five consecutive samples) in the visual control column than on a typical control chart. The basic lesson is that it is relatively easy to build simple but effective visual controls into a spreadsheet.

Chapter 4 will treat process capability indices in detail, but the next section will introduce some basic concepts.

Process Capability and Process Performance

Process capability indices and process performance indices are similar in concept but the difference lies in the method of calculation. The Automotive Industry Action Group (2005, 131–134) is an authoritative reference.

- Process *capability* indices rely on an estimate of the process standard deviation from subgroups:

$$\hat{\sigma}_C = \frac{1}{m} \sum_{i=1}^{m} \frac{R_i}{d_{2,i}} \quad \text{or} \quad \hat{\sigma}_C = \frac{1}{m} \sum_{i=1}^{m} \frac{s_i}{c_{4,i}}$$

where R_i is the range of the *i*th sample, s_i is the standard deviation of the *i*th sample, and $d_{2,i}$ and $c_{4,i}$ are control chart factors for the *i*th sample size. If all the samples are the same size, the average range or average sample standard deviation may be divided by the control chart factor for the given sample size. The C subscript for sigma means estimation from subgroups by means of control chart factors. This is the standard approach to setup of a control chart for variables.

- Process *performance* indices rely on an estimate of the process standard deviation from the individual measurements:

$$\hat{\sigma}_P = \sqrt{\frac{1}{N-1} \sum_{i=1}^{N} (x_i - \bar{x})^2}$$

TABLE 1.8

Capability and Process Performance Indices, Normal Data

Capability Indices	
$C_p = \dfrac{USL - LSL}{6\sigma_C}$	The C subscript for sigma means estimation from subgroups by means of control chart factors.
$CPL = \dfrac{\mu - LSL}{3\sigma_C}$	CPL reflects the process's ability to meet the lower specification.
$CPU = \dfrac{USL - \mu}{3\sigma_C}$	CPU reflects the process's ability to meet the upper specification.
$C_{pk} = \min[CPL, CPU]$	C_{pk} is the minimum of CPL and CPU. If the process is centered on the nominal, all four indices will be equal, which results in the least possible nonconforming product.
Performance Indices	
$P_p = \dfrac{USL - LSL}{6\sigma_P}$	The P subscript for sigma means estimation from the individual measurements as opposed to subgroup statistics.
$PPL = \dfrac{\mu - LSL}{3\sigma_P}$	
$PPU = \dfrac{USL - \mu}{3\sigma_P}$	
$P_{pk} = \min[PPL, PPU]$	

for N data. It differs from the previous estimate because it will include both within-subgroup and between-subgroup variation if the latter is present (e.g., nested variation sources). The maximum likelihood methods that fit nonnormal distributions to data rely on individual measurements, so process performance indices should be reported for those distributions.

Table 1.8 shows the capability and performance indices for data that follow the normal distribution.

A process with a performance or capability index of less than 1.33 (four standard deviations between the nominal and the specification limit) is generally not considered capable. A capability index of 2 (six standard deviations between the nominal and the specification limit) corresponds to Six Sigma quality, and in the absence of a shift in process mean, it corresponds to one defect or nonconformance per billion pieces at each specification limit.

The importance of the next section cannot be overemphasized. It is quite easy to put process data into statistical packages that assume normality, and then get clearly inappropriate results for nonnormal data. Excessive false alarms that undermine the workers' confidence in the SPC system are a frequent result, and Aesop's fable about the boy who cried "Wolf!" too many times is highly instructive. Inaccurate capability reports to vendors are likely to be another result. It is therefore impossible to overemphasize the importance of goodness-of-fit tests for the selected statistical distribution, and the tests work for nonnormal as well as normal data.

Goodness-of-Fit Tests

It is a basic tenet of statistical process control that a stable process is the basis of a valid control chart. This is why it is important to make control charts of the data that will serve as the basis for the charts on the shop floor, but it is also important to verify that the data fit the assumed distribution. If the fit is extremely poor, the preliminary control charts will probably, but not certainly, be unacceptable. If sample sizes are sufficiently large, in fact, the chart for process averages will look acceptable because of the central limit theorem. This is why it is also important to perform the goodness-of-fit tests.

This book recommends the quantile-quantile (or Q-Q) plot and the chi square goodness-of-fit tests. Both provide visual as well as quantitative outputs that allow the user to test the assumption that the selected statistical distribution really fits the process data. Other tests such as the Kolmogorov-Smirnov and the Anderson-Darling are available. Our experience is, however, that by the time the fit is bad enough to fail the Kolmogorov-Smirnov test, the problem has already become very obvious on the histogram and the quantile-quantile plot.

It is always important to remember that none of these tests proves that the distribution in question fits the data, because one can never prove the null hypothesis. A statement that "the fit to the normal distribution passes the chi square test" means there is not enough evidence to say that the normal distribution is not a good fit for the data. This does not prove, however, that the data are in fact normally distributed. Two or more distributions often deliver acceptable quantile-quantile plots and chi square tests, not to mention Kolmogorov-Smirnov results. In the absence of evidence or experience to the contrary, assumption of normality is often the best approach. On the other hand, if normal and gamma distribution models for particle counts or chemical impurities pass goodness-of-fit tests, the gamma distribution is probably the better model because of the underlying scientific basis—random undesirable arrivals—for its application. The nature of the process should generally play a role in the selection of the distribution if all other things are equal.

Quantile-Quantile Plot for Goodness of Fit

A quantile-quantile plot is a simple plot of the *ordered data* $x_{(i)}$ against the $(i - 0.5)/n$ quantile of the fitted distribution. Ordered data means organization of n individual measurements from smallest to largest: $x_{(1)}, x_{(2)}, ..., x_{(n)}$.

Other quantiles are sometimes used, such as $(i - 0.375)/(n + \frac{1}{4})$ and $i/(n + 1)$, but the reasoning behind $(i - 0.5)/n$ is easiest to understand. Suppose that

10 measurements are ordered from smallest to largest. The smallest represents the 0 to 10th percentile, of which 0.05 is the midpoint, and so on. The $(i - 0.5)/n$ quantile is simply

$$\mu + \sigma \times z_{(i)} = \mu + \sigma \times \Phi^{-1} \left(\frac{i - 0.5}{n} \right)$$

for the familiar normal probability plot, where Φ^{-1} is the inverse standard normal cumulative density. The results should include a tight fit around the best fit line, and also a high correlation coefficient. A point that is far from the best fit line is likely to be an *outlier* or special cause observation.

A MathCAD example with 100 simulated data from the distribution $N(60,2)$ (normal distribution with mean of 60 and variance of 2, Goodness_Fit1 in the simulated data spreadsheet) appears in Figure 1.16, and it illustrates the required calculations. The *line* variable on the ordinate is the regression line around which the ordered data points should scatter randomly.

Tests for goodness of fit are often not performed, however, because the computational tools are simply unavailable. The ability to do the job with Microsoft Excel or other spreadsheets is therefore highly desirable, and it should preferably require a minimum amount of user effort.

NORMINV($(i - 0.5)/100$,mean,sigma) will compute the $(i - 0.5)/n$ quantile of 100 data where i is the cell for the index (1 through 100) of the ordered measurements, the mean is their average, and sigma is the standard deviation. Do *not* use Excel's built-in functions for AVERAGE and STDEV for more than 30 pieces of data. SUM(A6:A105)/100 calculates the average of the indicated 100 measurements, and {=(SUM((A6:A105−C6)^2)/99)^0.5} returns their standard deviation. (The data set is available in SIMULAT.XLS in worksheet Goodness_Fit1.)

Excel's built-in regression routine can then determine the slope, intercept, and correlation coefficient of the ordered measurements versus the $(i - 0.5)/n$ quantiles. The Visual Basic for Applications function QQ_Plot does all of this automatically and provides three columns of output for use in an Excel scatter plot: the $(i - 0.5)/n$ quantiles as the axis, the sorted data to be plotted on the ordinate, and the best fit values for construction of the regression line. The function is entered on the spreadsheet as =QQ_Plot(A6:A105,"normal", 59.96235,2.124249,0,"C:\TEMP\QQ.txt") where the first argument is the range of data (empty cells are ignored), the second is the desired distribution (normal, sample variance, gamma, or Weibull; the program reads the first character and will understand lower or upper case), and the next three are the distribution parameters:

- Normal distribution: mean and standard deviation. The third parameter is ignored. In all cases, enter the best fit parameters for the dataset as opposed to hypothetical or standard parameters. As an example, if the process standard or nominal is 50.0 microns

n: = rows (x)   n = 100   i := 1..n   x_bar := mean(x)

$$\text{sigma} := \sqrt{\frac{1}{n-1} \cdot \sum_i (x_i - x\_bar)^2} \qquad \begin{bmatrix} x\_bar \\ sigma \end{bmatrix} = \begin{bmatrix} 59.962 \\ 2.124 \end{bmatrix}$$

x_ordered := sort(x)   Orders data from smallest to largest

$$\text{axis}_i := qnorm\left(\frac{i-0.5}{n}, x\_bar, sigma\right) \quad \text{qnorm is the inverse cumulative normal function}$$

$$\begin{bmatrix} m \\ b \\ r \end{bmatrix} := \begin{bmatrix} \text{slope (x\_ordered, axis)} \\ \text{intercept(x\_ordered, axis)} \\ \text{corr(x\_ordered, axis)} \end{bmatrix} \qquad \begin{bmatrix} m \\ b \\ r \end{bmatrix} = \begin{bmatrix} 0.994 \\ 0.368 \\ 0.995 \end{bmatrix} \quad \text{Best fit}$$

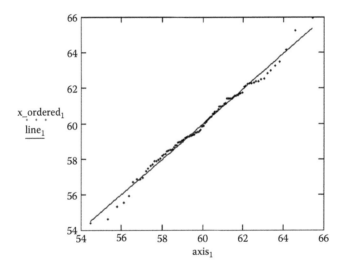

FIGURE 1.16
MathCAD normal probability plot.

while the grand average of the data is 50.2, use 50.2. The idea is to determine whether the distribution is appropriate for the dataset, as opposed to a hypothesis test on parameters.

- Sample variance from a normal distribution: sample size, and the second and third parameters are ignored.
- Gamma distribution: shape (α), scale (γ), and threshold (δ) parameters
- Weibull distribution: shape (β), scale (θ), and threshold (δ) parameters

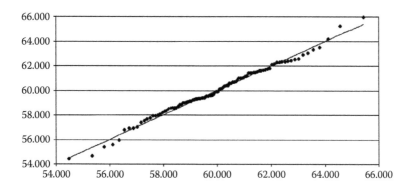

FIGURE 1.17
Excel normal probability plot using output from QQ_Plot.bas.

Visual Basic for Applications (VBA) functions apparently cannot, unlike compiled Visual Basic applications, alter the content of any spreadsheet cell other than the active one. The last entry is therefore the location of the text file for which the function is to write its report, the contents of which may then be pasted into Excel for use in a scatter diagram. Use the Data:Text to Columns procedure to convert the contents of the text file into columns, and then use the Chart Wizard to create the scatter diagram. Figure 1.17 shows the result for the 100 simulated data from the distribution $N(60,2)$.

QQ_Plot obtains slope 0.9966, intercept 0.206, and correlation $r = 0.9952$ as do Excel's regression routine and StatGraphics Centurion. The ordered measurements should be displayed as points only, and the best-fit regression line as a line only. The user may also have to format the chart's axis and ordinate to maximize use of the space.

Quantile-Quantile Plot for Sample Variance

Quantile-quantile plots also apply to sample variances from a normal distribution and, as shown in Chapter 3, sample ranges. When the sample size is n and the hypothetical standard deviation is σ_0,

$$\frac{(n-1)s^2}{\sigma_0^2}$$

should conform a chi square distribution with $n - 1$ degrees of freedom. This means that a plot of $(n - 1)s^2$ versus the $(i - 0.5)/N$ quantile of the chi square distribution for N samples should have a slope roughly equal to the population variance and an intercept of close to zero. Figure 1.18 shows this plot for 100 samples of 4 from the distribution $N(80,4)$. (The entire dataset is available in SIMULAT.XLS in the worksheet Goodness_Fit2.)

It is also possible to do a quantile-quantile plot in Microsoft Excel. In the case of $(n - 1)s^2$ versus chi square, use STDEV to calculate the sample

$N := \text{rows}(s)$ $i := 1..N$

$y_i := 3 \cdot (s_i)^2$ $y := \text{sort}(y)$ Ordered variances times degrees of freedom

$x_i := \text{qchisq}\left[\left(\dfrac{i-0.5}{N}\right), 3\right]$ (i-0.5)/N quantile, chi square with 3 degrees of freedom

$\begin{bmatrix} b_0 \\ b_1 \\ \text{corr} \end{bmatrix} := \begin{bmatrix} \text{intercept}(x,y) \\ \text{slope}(x,y) \\ \text{corr}(x,y) \end{bmatrix}$ $\begin{bmatrix} b_0 \\ b_1 \\ \text{corr} \end{bmatrix} = \begin{bmatrix} 1.349 \\ 3.825 \\ 0.997 \end{bmatrix}$ 100 samples of 4 simulated from N(80,4); the slope should be about 4

3*sample variance

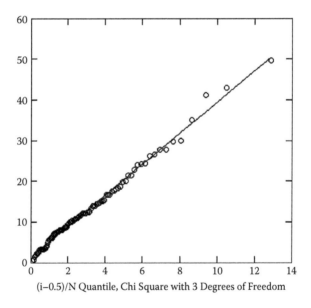

FIGURE 1.18
$(n-1)s^2$ versus χ^2 with 3 degrees of freedom.

standard deviation, square it to get the sample variance, and then multiply by $(n - 1) = 3$. The SORT procedure under the DATA menu will order the results from smallest to largest. =CHIINV((101–H6–0.5)/100,3) returns the *quantile* of chi square for (H6 − 0.5)/100, noting that the first argument of CHIINV is the upper tail area, or 1 minus the cumulative density. As an example, CHIINV(0.05,3) returns the 0.95 quantile for the distribution with 3 degrees of freedom, and 0.05 is the upper tail area.

TABLE 1.9

Data for Quantile-Quantile Plot in
Microsoft Excel

i	Chi Square	Sorted 3*s²	Best Fit
1	0.072	0.583	1.623
2	0.152	0.735	1.929
3	0.216	1.664	2.174
4	0.273	2.050	2.394
5	0.326	2.384	2.597
6	0.377	2.517	2.790

Use of the built-in regression routine (**Regression** under **Tools:Data Analysis**) returns the same slope, intercept, and correlation coefficient that appear above. The slope and intercept can in turn generate the best fit line as shown in Table 1.9; the best fit value is the $(i - 0.5)/n$ quantile of the chi square distribution times the fitted slope plus the intercept.

Use of Excel's scatter chart with the sorted $3s^2$ values plotted as points with the best fit values as a line results in Figure 1.19.

The Visual Basic for Applications function QQ_Plot can perform all of this automatically except for generation of the plot itself. =QQ_Plot(F6:F105,"Sample variance",4,0,0,"C:\TEMP\QQ_Variance.txt") returns the same slope, intercept, and correlation (3.825, 1.349, and 0.9967) as the previous methods along with columns of axis values, ordered $(n - 1)s^2$ values, and the best fit regression line.

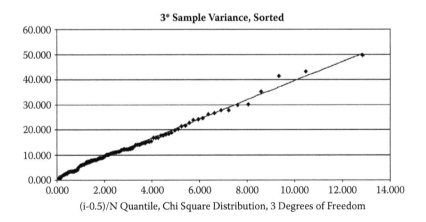

FIGURE 1.19
$(n-1)s^2$ versus χ^2 with 3 degrees of freedom, Microsoft Excel.

Chi Square Test

A histogram is a visual tool that shows how well the raw data conform to the selected distribution. The judgment that the histogram "looks good" or "looks bad" is, however, subjective, and the chi square test adds a quantitative measurement. The procedure is as follows:

1. Prepare a histogram of the data.
   - For N measurements, use $N^{0.5}$ cells (Messina, 1987, 16), or $4(0.75)$ $(n-1)^2)^{0.2}$ cells (Shapiro, 1986, 24).
   - N should be no less than 30, and it should preferably be 100 or more (Juran and Gryna, 1988, 23.72). The need for enough measurements to define the distribution parameters and allow testing for goodness of fit cannot be overemphasized.
   - One problem with the chi square test is that, at least in borderline situations, a passing or failing result will depend on the number of cells. In the case of the simulated 2-parameter Weibull distribution (Chapter 2), an initial selection of 15 cells for 200 data yields a P value of 0.65 in StatGraphics, while 16 cells deliver a P value of 0.08—almost low enough to fail at the traditional 0.05 significance level standard. The good quantile-quantile plot should, however, remind the user that the chi square test's results *do* depend on the number of cells, so a poor result for a given number of cells should not be interpreted as absolute proof that the distribution does not fit the data. We could also envision a situation in which the chi square test delivers a relatively good result while the quantile-quantile plot warns the practitioner that the distribution is clearly inappropriate. The computer is a powerful aid to the judgment of the quality professional, but it should never become a substitute for it.

2. Let f_k be the number of observations in the kth out of m cells.

3. Compute the expected number of observations (E_k) in each cell, whose lower and upper boundaries are $[L_k, U_k]$. If $f(x)$ is the probability density function and $F(x)$ is the cumulative function, $E_k = N \int_{L_k}^{U_k} f(x)dx = N[F(U_k) - F(L_k)]$. In other words, E_k is the number of measurements times the area under the probability density function for each cell.
   - In the case of the normal distribution,

$$E_k = N \left[\Phi\left(\frac{U_k - \mu}{\sigma}\right) - \Phi\left(\frac{L_k - \mu}{\sigma}\right) \right].$$

   - For the cells at the ends, use the lower and upper tail areas, respectively. To check the results, make sure the total observed and expected counts equal the number of measurements.

- The underlying assumptions behind the test require that the expected count in each cell be no less than five. If this is not true, reduce the number of cells and/or combine tail cells until it is.

4. Compute the chi square test statistic

$$\chi^2 = \sum_{k=1}^{m} \frac{(f_k - E_k)^2}{E_k}.$$

- Note intuitively how this measures the badness of fit by squaring the difference between the observed and expected counts.

5. Reject the null hypothesis that the data fit the selected probability density function if, given m cells and the estimation of p parameters from the data, $\chi^2 > \chi^2_{m-1-p;\alpha}$ where α is the desired significance level (typically 5 percent).

- In the case of the normal distribution, two parameters (μ and σ) are estimated from the data so $p = 2$. If there are six cells, there are $6 - 2 - 1 = 3$ degrees of freedom. An exponential or Poisson distribution has only one parameter that is estimated from the data.

Example

Given 100 simulated data from the distribution $N(60,2)$ (the same dataset for the MathCAD normal probability plot, Goodness_Fit1). Statistical packages perform the chi square test automatically, but Microsoft Excel is useful for illustration of the procedure's details. The spreadsheet's histogram function yields the first part of Table 1.10. U refers to the upper limit of the cell; for example, if U is 58, the count is of all points that are between 57 and 58.

TABLE 1.10

Chi Square Test for 100 Normally Distributed Measurements

	E	F	G	H	I	J	K	L
	U	Bin	Frequency	E	U	f	E	χ^2
6	56	56	5	3.107				
7	57	57	4	5.051	57	9	8.158	0.087
8	58	58	7	9.622	58	7	9.622	0.715
9	59	59	15	14.746	59	15	14.746	0.004
10	60	60	21	18.181	60	21	18.181	0.437
11	61	61	16	18.032	61	16	18.032	0.229
12	62	62	15	14.388	62	15	14.388	0.026
13	63	63	11	9.236	63	11	9.236	0.337
14	64	64	3	4.769	More	6	7.636	0.351
15	65	65	1	1.981				
16		More	2	0.886				
17				100		100	100	2.186

Here is a sample calculation for the expected frequency in cell H7: =100*(NO RMDIST(F7,C$6,C$7,1)−NORMDIST(F6,C$6,C$7,1)) where C$6 is the data average and C$7 is the standard deviation of the data. Remember that STDEV cannot compute these for more than 30 points; the user must program the spreadsheet to calculate

$$\bar{x} = \frac{1}{N-1}\sum_{i=1}^{N} x_i \quad \text{and} \quad \sigma = \sqrt{\frac{1}{N-1}\sum_{i=1}^{N}(x_i - \bar{x})^2}.$$

Since the expected count in each cell should be 5 or more, the tail cells are combined accordingly in columns I through L. As a check on the bookkeeping, note that the observed and expected counts add up to 100.

The chi square result for cell L7 is =(J7-K7)^2/K7, and the others are calculated similarly. The result of 2.186 has $8 - 1 - 2 = 5$ degrees of freedom. Comparison with the tabulated value of 11.07 for five degrees of freedom and $\alpha = 0.05$ shows that the hypothesis that the distribution is normal cannot be rejected at the 5% significance level.

A Visual Basic for Applications function, ChiSquare_Normal, automates this entire process. The arguments of the function are

- Data: the user-selected range of cells for which the test is to be performed
- N_Cells: the number of cells into which the data are to be sorted
- Mean: the mean of the hypothetical normal distribution
- Sigma: the standard deviation of the hypothetical normal distribution
- OutPut_File: the name of the text file (including directory) into which the results are to be written.

Table 1.11 shows the results for the simulated normal distribution $N(60,4)$ and ten cells. The results are stored in C:\TEMP\ChiSquare.txt and the chi square test statistic appears in the cell into which the function =ChiSquare_ Normal(A6:A105,10,59.96,2.124,"C:\TEMP\ChiSquare.txt") was entered. Note that the sums of the observed and expected counts are presented to confirm that all data have been accounted for.

The user could then copy the results into an Excel spreadsheet (Figure 1.20) and merge the first two and last three cells manually, or else rerun the function with a different number of cells.

StatGraphics Centurion 15 yields results (Table 1.12) similar to those in Table 1.9, but it does not automatically combine the cells at each end of the distribution.

The StatGraphics report adds, "Chi-Squared = 3.78237 with 7 d.f. P-Value = 0.804475," where the P-value is the significance level at which the hypothesis of a good fit is rejectable given the chi square test statistic. In other words, if the cutoff is given at 3.78 instead of 14.067 (chi square for 7 degrees of freedom

TABLE 1.11

Results from VBA Chi Square Test Function

Chi square test for goodness of fit, normal distribution
VBA.xls
Goodness_of_Fit_Normal
100 Data, 10 Cells
Mean 5.9960E+01 Standard deviation 2.1240E+00

Cell	Floor	Ceiling	Observed	Expected	ChiSquare
1	5.441E+01	5.557E+01	3	1.926	.598
2	5.557E+01	5.672E+01	2	4.434	1.336
3	5.672E+01	5.788E+01	8	9.962	.387
4	5.788E+01	5.903E+01	18	16.770	.090
5	5.903E+01	6.019E+01	24	21.152	.384
6	6.019E+01	6.134E+01	16	19.988	.796
7	6.134E+01	6.250E+01	20	14.153	2.416
8	6.250E+01	6.365E+01	6	7.508	.303
9	6.365E+01	6.481E+01	1	2.983	1.319
10	6.481E+01	6.596E+01	2	1.123	.684
			100	100.00	8.311

and an upper tail of $\alpha = 0.05$), the test has an 80.4% chance to reject a normal distribution as nonnormal. Figure 1.21 shows how to interpret the P value.

Figure 1.22 shows the histogram from StatGraphics Centurion. This is where the user can make a subjective judgment as to whether the fit is a good one, but it is important to remember that the chi square test is quantitative.

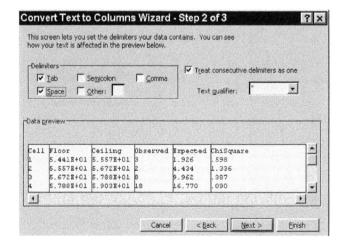

FIGURE 1.20
How to copy text file contents to Excel.

TABLE 1.12

Chi Square Test from StatGraphics Centurion

	Lower Limit	Upper Limit	Observed Frequency	Expected Frequency	Chi Squared
At or below		56.0	5	3.11	1.15
	56.0	57.0	4	5.05	0.22
	57.0	58.0	7	9.62	0.71
	58.0	59.0	15	14.75	0.00
	59.0	60.0	21	18.18	0.44
	60.0	61.0	16	18.03	0.23
	61.0	62.0	15	14.39	0.03
	62.0	63.0	11	9.24	0.34
	63.0	64.0	3	4.77	0.66
Above	64.0		3	2.87	0.01

Figure 1.23 is the quantile-quantile plot from StatGraphics Centurion (note the similarity to the plot from MathCAD). This is what the Q-Q plot should look like if the distribution is a good fit.

Figure 1.24 is the quantile-quantile plot as prepared by Minitab 15. In this case, the ordered data appear on the horizontal axis while the inverse

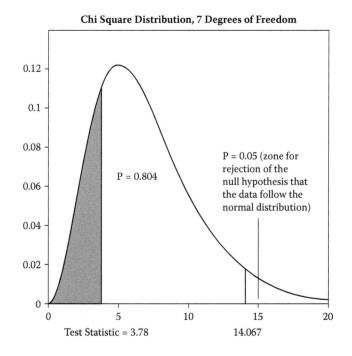

Chi Square Distribution, 7 Degrees of Freedom

P = 0.804

P = 0.05 (zone for rejection of the null hypothesis that the data follow the normal distribution)

Test Statistic = 3.78   14.067

FIGURE 1.21
P value for hypothesis testing.

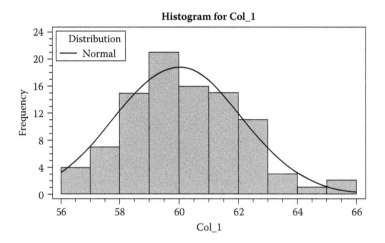

FIGURE 1.22
Histogram of data from a normal distribution.

cumulative distribution of $(i - 0.5)/n$ or a similar estimate of the quantile midpoint is on the ordinate. The format of the latter is that of the familiar normal probability plot. The P-Value is the confidence with which the hypothesis of a normal distribution may be rejected (significance level).

This section has covered the quantile-quantile plot and the chi square test, both of which should be performed to assess the goodness of fit between the selected distribution and the data.

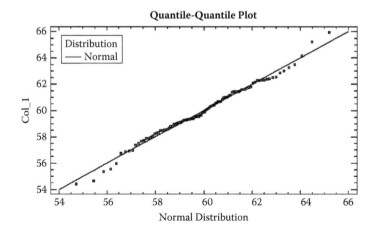

FIGURE 1.23
Normal probability plot, StatGraphics Centurion.

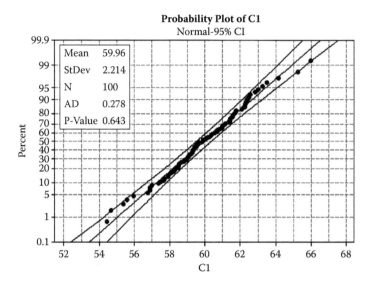

FIGURE 1.24
Normal probability plot, Minitab.

Although attribute control charts are not this book's primary focus, it does stress the virtues of exact control limits as opposed to approximations. The traditional limits of the charts for sample range and sample standard deviation are only approximations because neither of these sample statistics follows a normal distribution. The same applies to the traditional limits of the charts for attributes, and the next section offers a significant improvement.

Multiple Attribute Control Charts

The traditional charts for attributes—number nonconforming, nonconforming fraction, defect count, and defect density—were developed for convenience in hand calculation and manual plotting. The computing power of modern spreadsheets allows deployment of charts that do not rely on approximations, and that can give immediate visual signals when a point exceeds its control limits.

Levinson (1994a, 2004) introduces the *multiple attribute control chart*, which is essentially a check sheet or tally sheet with control limits. It is capable of tracking multiple defect or nonconformance causes on a single spreadsheet, and it does not require a consistent sample size. The multiple attribute chart is in turn an extension of the control chart that was introduced by

Cooper and Demos (1991) in which colored symbols indicate out-of-control measurements.

The problem with the traditional control charts for attributes is that they rely on the normal approximations for the binomial (number nonconforming) and Poisson (defects or nonconformities) distributions:

Equation Set 1.5: Traditional Attribute Control Charts

np (number nonconforming)	$np_0 \pm 3\sqrt{np_0(1-p_0)}$ where $\sqrt{np_0(1-p_0)}$ is the standard deviation of the binomial distribution and p_0 is the historical nonconforming fraction
p (fraction nonconforming)	$p_0 \pm 3\sqrt{\dfrac{p_0(1-p_0)}{n}}$
c (defect count)	$c_0 \pm 3\sqrt{c_0}$ noting that $\sqrt{c_0}$ is the standard deviation of the Poisson distribution, where c_0 is the historical nonconforming fraction.
u (defect density)	$u_0 \pm 3\sqrt{\dfrac{u_0}{n}}$ where $u = \dfrac{c}{n}$

These approximations rely on the assumption that the distribution mean is at least 4 if not higher. In other words, the conditions under which the traditional control limits are even marginally valid involve four or more rejects or defects per sample. These conditions are not consistent with even a minimally acceptable process capability of 1.33 (63 defects per million opportunities if the process is centered on the nominal).

It is, however, possible to program a spreadsheet to use the exact quantiles of the binomial or Poisson distribution to set the control limits, and the procedure is extremely simple as shown in Table 1.13.

Cell C2 contains the desired bilateral false alarm risk. Row 4 contains the types of nonconformances, and Row 5 the standard or historical nonconformance rate. The user then enters the sample size and the number of each type of nonconformance for each lot, as shown for ten lots. The data entry format is simply that of the familiar check sheet or tally sheet, which is perhaps the most basic of the seven traditional quality improvement tools.[4] Row 16 shows how the user can sum the columns at any time to make a Pareto chart of the nonconformance causes.

The visual control is built into the conditional formatting of each cell as shown in Figure 1.25 for cell C6.

Both conditional formats check to make sure a number is present (ISNUMBER function) before they perform the indicated test, so that a blank is not treated as a zero. The cell will turn green if the entry is below the lower

TABLE 1.13

Multiple Attribute Control Chart, Binomial Distribution

	A	B	C	D	E	F
1	Multiple Attribute Control Chart (binomial)					
2	False alarm risk		2.00%			
3						
4	Nonconformance		A	B	C	D
5	Lot	Sample	0.50%	1.00%	0.10%	2%
6	1	600	3	5	0	16
7	2	600	5	13	1	21
8	3	600	2	7	1	18
9	4	600	7	9	3	15
10	5	1000	5	12	0	30
11	6	1000	9	17	3	31
12	7	1000	5	8	2	16
13	8	1000	5	6	1	15
14	9	1000	4	6	3	11
15	10	100	0	0	0	1
16		7500	45	83	14	174

control limit and red if it is above the upper control limit. Given a noncon-formance count of x (C6), a sample size of n ($B6), a historical or standard nonconforming fraction p_0 (C$5), and a two-sided false alarm risk α ($C$2), x is below the lower control limit if

$$\sum_{k=0}^{x} \frac{n!}{k!(n-k)!} p_0^x (1-p_0)^{n-k} < \frac{\alpha}{2}.$$

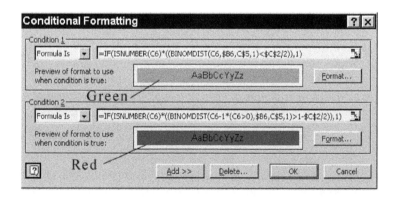

FIGURE 1.25
Visual control, multiple attribute control chart.

The upper control limit is defined as

$$\sum_{k=0}^{UCL} \frac{n!}{k!(n-k)!} p_0^x (1-p_0)^{n-k} \geq 1 - \frac{\alpha}{2},$$

so the test

$$\sum_{k=0}^{x-1} \frac{n!}{k!(n-k)!} p_0^x (1-p_0)^{n-k} > 1 - \frac{\alpha}{2}$$

is equivalent to the test "x>UCL." The 1*(C>0) factor prevents an attempt to find the cumulative binomial distribution for −1 if x happens to be zero.

The random simulation produced an out-of-control signal in cell D7, with 13 nonconformances out of 600 pieces and $p_0 = 0.01$. BINOMDIST(12,600,0.01,1) yields a cumulative probability of 0.9915, which makes 12 the upper control limit for this sample size. It is important to note that, even though 12 is the 0.9915 quantile, it is not outside the control limit for a one-sided false alarm risk of 1 percent. Eleven is the 0.980 quantile, so the chance of getting 12 or more nonconformances is almost 2 percent. Twelve is therefore the lowest UCL for which the false alarm risk meets the criterion "1 percent or less."

This section has illustrated the ability of spreadsheets to change the color of a cell in response to an out-of-control condition. This important visual control will work for other charts as well, and the production operator or inspector does not even have to look at the actual control chart to determine the status of a measurement or a sample.

If p_0 is extremely small, however, a small sample size may result in an out-of-control signal at the *upper* control limit because the cumulative binomial or Poisson probability for zero exceeds 1 − α/2. As a simple example, the cumulative binomial probability of zero for a sample of 10 and a nonconforming fraction of 0.001 is slightly more than 0.99, which means the selected false alarm risk of 1 percent in each tail would result in an out-of-control signal. This is not something that should arise in actual practice, but it can be avoided by adding the requirement "C6>0" (one or more defects or nonconformances must be present) to the conditional test. An exponentially weighted moving average (EWMA) chart may be helpful when p_0 is very small, and the exponential distribution itself is a viable model for parts produced between defects.

Exponentially Weighted Moving Average Chart for Defects

When p_0 is extremely small, an EWMA chart can track time, or presumably number of items produced, between defects or nonconformances. Let z_0 be the initial measurement in the series while λ is a constant weighting factor between

0 and 1. Montgomery (1991, 306) recommends that λ be between 0.05 and 0.25. If t_i is the time between the $(i-1)$th and ith occurrence (such as a defect),

Equation Set 1.6

$$z_1 = \lambda t_1 + (1-\lambda)t_0$$

$$z_i = \lambda t_i + (1-\lambda)z_{i-1}$$

$$UCL = \bar{t} + 3\bar{t}\sqrt{\frac{\lambda}{2-\lambda}}$$

$$LCL = \bar{t} - 3\bar{t}\sqrt{\frac{\lambda}{2-\lambda}} \quad \text{but not less than 0}$$

The Automotive Industry Action Group (2005, 111–113) also discusses EWMA charts and notes their utility in the detection of small process shifts with slowly drifting means. Note that "time" can consist of the number of units produced as opposed to actual time, just as "time" in reliability statistics covers numerous measurements such as miles driven, fatigue failure cycles, and other characteristics.

Use of the Exponential Distribution to Track Defects or Nonconformances

The Poisson distribution's parameter is events per unit time or events per N items produced and is in fact the hazard function of the exponential distribution itself. This suggests that the number of items produced between defects or nonconformances can be modeled by the exponential distribution.

The usual metric for defects or nonconformances is the count per number of parts produced. Suppose the following, however:

1. The event rate is at the defects per million opportunities level. This makes the Poisson model valid for nonconformances as well as defects.

2. The process makes hundreds of thousands of parts.

3. The process operates like a sequential assembly line to the extent that it is possible to report the number of *pieces produced between defects or nonconformances*.

The geometric distribution is a common model for discovery sampling, or estimation of the number of items one must inspect to find a single nonconformance. It can therefore be used to simulate the number of pieces needed to get one defect. If U is a random number from the uniform distribution on interval [0,1], simulation is as follows:

$$f(n) = (1-p)^n p \quad \text{and} \quad F(n) = 1-(1-p)^n \Rightarrow$$

$$n = \frac{\ln(1-F(n))}{\ln(1-p)} = \frac{\ln U}{\ln(1-p)}$$

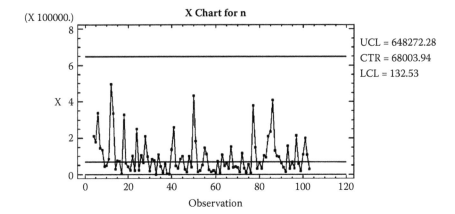

FIGURE 1.26
Parts produced between defects (exponential distribution).

Consider, for example, a process that generates, on average, 10 defects per million opportunities. StatGraphics fits 100 simulated data (Exponential_ Defects in the simulated data worksheet), in terms of parts produced between defects, to an exponential distribution with a mean close to the reciprocal of the hazard rate of 10 DPMO: 98108.9 pieces between defects. This would be the mean time between failure (MTBF) in reliability statistics. As shown in Figure 1.26, StatGraphics can plot this data on an exponential distribution control chart with the following control limits and center line:

$$\text{Exponential Distribution CDF } F(x) = 1 - \exp\left(-\frac{x}{\theta}\right) \Rightarrow x = -\theta \ln(1 - F(x))$$

UCL $x(0.99835) = -98108.9 \ln(1 - 0.99865) = 648269$

CL $x(0.5) = -98108.9 \ln(0.5) = 68004$

LCL $x(0.00135) = -98108.9 \ln(1 - 0.00135) = 133.5$

StatGraphics, which carries all the significant figures, gets a slightly higher UCL but it is actually the LCL that is of interest. If there are 132 or fewer pieces between defects, this is strong evidence that the defect rate has increased.

This chapter has described the traditional control charts that assume that the underlying process follows a normal distribution. The next chapter will introduce nonnormal distributions that are found in many manufacturing applications, and then it will show how to construct appropriate control charts.

Exercises

Exercise 1.1

A supplier provides a capability report for a chemical product with an upper specification limit of 16 ppm impurities. The report cites 50 samples of 10 with a grand average of 6.077 and a standard deviation of 1.818 based on subgroup standard deviations, from which the supplier calculates an upper process capability index of $(16 - 6.077)/(3 \times 1.818) = 1.819$. This implies about 24 nonconformances per billion. The report also includes the 50 sample averages and standard deviations in EXERCISES.XLS (1_1 tab), but not the individual measurements that also appear on this worksheet. The averages appear to follow a normal distribution as shown here:

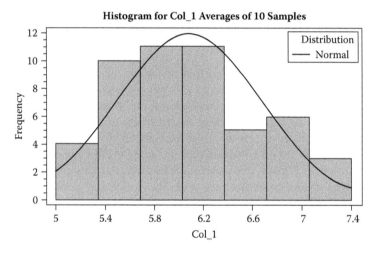

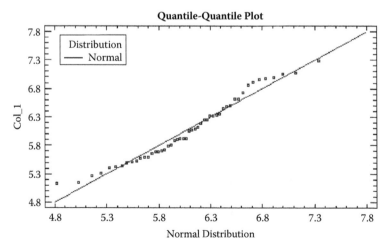

Should the customer accept the supplier's statement that the upper process capability estimate is 1.819?

Exercise 1.2

Given a process with a historical nonconforming fraction of 1 percent, what is the traditional upper control limit for a sample of 400 on an np (number nonconforming) chart? What is the actual false alarm risk for this control limit?

Exercise 1.3

What is the upper control limit for an s chart with a sample size of 5 and a historical standard deviation of 1.0 mil? What is the actual false alarm risk for this chart?

Exercise 1.4

Which of the following are likely to be rational subgroups?

a. Five parts selected at random from a batch of 500 from a heat treatment furnace

b. Four pieces selected at random or systematically (e.g., every tenth part) from an inline heat treatment furnace that processes one piece at a time

c. Photoresist thicknesses on five silicon wafers from a spin coater

d. Metal thicknesses on three silicon wafers selected at random from a batch of 20 from a metallization tool

e. Hole diameters from five pieces selected randomly or systematically from the output of a drill press

Exercise 1.5

Six Sigma often quotes "3.4 defects per million opportunities," which corresponds to a 1.5 sigma shift in the process mean. What is the X (individuals) chart's power to detect this process shift, and what is the average run length?

Solutions

Exercise 1.1

The first hint that there might be a problem with the supplier's analysis is the systematic pattern on the quantile-quantile plot. The points should scatter randomly around the best fit line. In addition, the fact that these are averages of ten measurements means that their conformance to a normal distribution is no guarantee that the underlying measurements come from a normal distribution. The central limit theorem says that the sample average from any distribution will approach normality with a large enough sample size. A plot of ordered sample variances s^2 times $(n-1)$ versus the $(i-0.5)/N$ quantile of the chi square distribution for $N = 50$ samples and with 9 degrees of freedom is highly instructive:

$N := \text{rows}(s)$ $i := 1 .. N$

$y_i := 9 \cdot (s_i)^2$ $y := \text{sort}(y)$ Ordered variances times degrees of freedom

$x_i := \text{qchisq}\left[\left(\dfrac{i-0.5}{N}\right),9\right]$ (i-0.5)/N quantile, chi square with 3 degrees of freedom

$\begin{bmatrix} b_0 \\ b_1 \\ \text{corr} \end{bmatrix} := \begin{bmatrix} \text{intercept}(x,y) \\ \text{slope}(x,y) \\ \text{corr}(x,y) \end{bmatrix}$ $\begin{bmatrix} b_0 \\ b_1 \\ \text{corr} \end{bmatrix} := \begin{bmatrix} -9.592 \\ 4.473 \\ 0.967 \end{bmatrix}$ 50 samples of 10

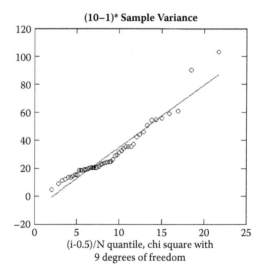

(10−1)* Sample Variance

(i-0.5)/N quantile, chi square with
9 degrees of freedom

Although the pattern of the points in the quantile-quantile plot for sample averages is merely suspicious, that for the sample variances shows a definite problem.

The truth is that the data were simulated from a 3-parameter gamma distribution, as is common for applications of this nature, with $\alpha = 3$, $\gamma = 1$, and $\delta = 3$. The actual upper process performance index, as calculated from the

500 individual measurements, is not 1.819 with a 24-parts per billion (ppb) nonconformance rate. It is 1.10, which corresponds to 456 defects per million opportunities, and it does not even meet the generally accepted standard for a capable process. The next chapter will show how to fit this distribution to the raw data, but the key lesson is that assumption of normality for a non-normal process can result in estimates of nonconforming fractions that are off by more than four orders of magnitude.

Exercise 1.2

For a sample of 400 and a historical nonconformance rate of 1 percent, $np_0 + 3\sqrt{np_0(1-p_0)} = 4 + 3\sqrt{4 \times 0.99} = 9.97$. Ten or more nonconformances would be out of control.

The actual false alarm risk is the chance of getting 10 or more nonconformances:

$$1 - \sum_{x=0}^{9} \frac{400!}{x!(400-x)!} 0.01^x \times 0.99^{400-x} = 0.0078$$

(Recursive formulas are available to avoid the need to calculate 400!, which would result in computational overflow.) Also, noting the large sample and relatively small chance of occurrence, the Poisson approximation to the binomial distribution could be used; the cumulative Poisson probability for 9 events given a Poisson mean of 4 is 0.992, which yields the same result. In Microsoft Excel, =1−BINOMDIST(9,400,0.01,1) returns 0.0078. The result is more than five times the implied risk of 0.00135 for Shewhart control limits.

Exercise 1.3

The upper control limit for the s chart when the standard deviation is given (theoretical control chart) is $B_6\sigma_0$. B_6 for a sample of 4 is 2.088, so the upper control limit is 2.088 times 1.0 mil = 2.088 mils. The actual false alarm risk can be determined from the chi square distribution:

$$\chi^2 = \frac{(n-1)s^2}{\sigma_{0,k}^2} = \frac{3 \times (2.088 \ \text{mil})^2}{1 \ \text{mil}^2} = 13.08.$$

In Microsoft Excel, CHIDIST(13.08,3) returns a false alarm risk of 0.00447, or more than three times the implied 0.00135 false alarm risk of a Shewhart chart.

Exercise 1.4

Items b, c, and e are rational subgroups because the tool or process unit (single-unit heat treatment furnace, spin coater, or drill press) is the only source of variation.

Items a and d are unlikely to be rational subgroups because the equipment (batch heat treatment furnace, batch metallization tool) has between-batch and within-batch variation. This can in fact be determined through one-way analysis of variance, in which each batch is considered a separate treatment, or through a components of variance analysis. Chapter 4 discusses this further.

Exercise 1.5

The standard normal deviate may be easily used to find the power of an X chart to detect a 1.5 sigma change in the process mean. Suppose this is an increase (a similar approach works for a decrease).

$$z = \frac{UCL - \mu}{\frac{\sigma}{\sqrt{n}}} = \frac{(\mu_0 + 3\sigma) - (\mu_0 + 1.5\sigma)}{\sigma} = 1.5 \quad \Phi(1.5) = 0.933193$$

This is the chance that any observed individual will be below the upper control limit, so the power is the 6.681 percent chance that the measurement will be above the UCL. The average run length is the reciprocal of the power, or 14.97. It will require on average about fifteen measurements to detect the process shift.

Endnotes

1. This is in fact an autoregressive AR(1) time series model in which $X_t = a + \phi X_{t-1} + \varepsilon$ where ε is the random variation, which is in this case a random number from the normal distribution with mean zero and variance 1. Minitab fits these data to an AR(1) model with $a = 201.1$ and $\phi = -1.01$.
2. The ISO 9000 document hierarchy consists of the quality manual (first tier, at the top), second-tier documents that span many departments or activities (such as the document control procedure for the entire factory), third-tier work instructions for individual jobs, and fourth-tier documents such as quality records.
3. Originally the *Western Electric Handbook*.
4. The seven traditional quality improvement tools are the check sheet, histogram, Pareto chart, cause-and-effect diagram, process flowchart, SPC chart, and scatter diagram.

2

Nonnormal Distributions

This chapter will discuss common nonnormal distributions, construction of appropriate control charts, and calculation of meaningful process performance indices. The foundation of these methods is as follows:

1. Control charts should behave like traditional Shewhart charts in terms of false alarm risks. Since the false alarm risk for each tail of a Shewhart chart with 3-sigma limits is 0.00135, the risk should be the same for the nonnormal chart at each control limit.

   - The chart's power to detect process shifts (e.g., average run length) is unlikely to match that of a corresponding Shewhart chart for normal data, but it can be calculated for any given change in the distribution parameters.

2. Process performance indices must reflect the nonconforming fraction accurately.

3. *Always* perform goodness-of-fit tests to verify that the selected distribution is a good fit for the data!

In some cases, transformation of the nonnormal distribution to a normal distribution will simplify matters enormously for shop floor control charts, but it is important to note that such transformations may lose accuracy in the regions of interest for the process performance index.

Transformations

It is often possible to transform one probability density function into another. The procedure for a single variable is as follows. Given a probability density function $f(x)$ and a 1:1 relationship between variables x and y, the objective is to define a corresponding density function $g(y)$. Then

$$g(y) = f(x(y))J\left(\frac{x}{y}\right) \text{ where J is the Jacobian operator} :J\left(\frac{x}{y}\right) = \frac{\partial x}{\partial y}$$

The Jacobian must not be zero in the region of interest. Then:

1. Convert the limits of integration
2. Convert the integrand
3. Convert the argument

The procedure is as follows for a joint distribution. Given $f(x,y)$ and a 1:1 relationship between (x,y) and (u,v):

$$g(u,v) = f(x(u,v), y(u,v))J\left(\frac{x,y}{u,v}\right) \text{ where } J\left(\frac{x,y}{u,v}\right) = \begin{vmatrix} \dfrac{\partial x}{\partial u} & \dfrac{\partial x}{\partial v} \\ \dfrac{\partial y}{\partial u} & \dfrac{\partial y}{\partial v} \end{vmatrix} \quad \text{(Set 2.1)}$$

This allows determination of marginal probabilities for u and v, e.g., $h(u) = \int_{min}^{max} g(u,v)dv$ where min and max are the integration limits of v.

In summary, transformations allow the identification or development of probability density functions to handle transformed variables. The following section shows an example in which such a transformation can make measurements from a gamma distribution behave *almost* as if they come from a normal distribution, at least for control chart purposes. The transformation does not, however, yield an accurate process performance index.

Square Root Transformation for Random Arrivals

Levinson, Stensney, Webb, and Glahn (2001) demonstrate why the square root of undesirable random arrivals (particles in semiconductor equipment) should follow the normal distribution. Previous experience shows that the gamma distribution, of which the chi square distribution is a special case, is a good model for particle counts. When the mean of a chi square distribution is relatively large, Fisher's approximation can relate it to a normal distribution (Johnson and Kotz, 1970, 176). The words "when the mean is relatively large" and "approximation" suggest, however, that the resulting normal distribution is not an exact model for the process.

Fisher's approximation is

$$\chi_v^2 = \frac{1}{2}(z + \sqrt{2v-1})^2$$

where z is the standard normal deviate and v("nu") is the degrees of freedom. Then

$$\sqrt{\chi_v^2} = \sqrt{\frac{1}{2}(z + \sqrt{2v-1})} \Rightarrow \sqrt{\chi_v^2} = \frac{\sqrt{2}}{2}\sqrt{2v-1} + \frac{\sqrt{2}}{2}z$$

In other words, the square root of chi square with v degrees of freedom equals a constant plus a random number from a normal distribution. Noting that

$$x \sim N(\mu, \sigma^2) \Rightarrow cx \sim N(c\mu, c\sigma^2)$$

$$z \sim N(0,1) \Rightarrow \frac{\sqrt{2}}{2} z \sim N\left(0, \frac{1}{2}\right)$$

$$\sqrt{\chi_v^2} = \frac{\sqrt{2}}{2}\sqrt{2v-1} + \frac{\sqrt{2}}{2} z \sim N\left(\frac{\sqrt{2}}{2}\sqrt{2v-1}, \frac{1}{2}\right)$$

This shows that the square root of a response from this chi square distribution should in fact approximate a normal distribution, and the approximation should improve as v increases. The latter is not, however, desirable when the application involves random arrivals such as particles or impurities.

The next step is to relate the chi square distribution to the gamma distribution

$$f(x) = \frac{\gamma^\alpha}{\Gamma(\alpha)} x^{\alpha-1} \exp(-\gamma x).$$

Let $y = 2\gamma x$ and transform this to a chi square distribution.

Recalling that $g(y) = f(x(y)) J\left(\frac{x}{y}\right)$,

$$J\left(\frac{x}{y}\right) = \frac{\partial x}{\partial y} = \frac{1}{2\gamma}$$

$$g(y) = \frac{\gamma^\alpha}{\Gamma(\alpha)}\left(\frac{y}{2\gamma}\right)^{\alpha-1} \exp\left(-\frac{y}{2}\right)\frac{1}{2\gamma}$$

$$= \frac{\gamma^\alpha}{\Gamma(\alpha)}\left(\frac{1}{2\gamma}\right)^\alpha y^{\alpha-1} \exp\left(-\frac{y}{2}\right)$$

$$= \frac{1}{2^\alpha \Gamma(\alpha)} y^{\alpha-1} \exp\left(-\frac{y}{2}\right)$$

$$v = 2\alpha \Rightarrow g(y) = \frac{1}{2^{\frac{v}{2}}\Gamma\left(\frac{v}{2}\right)} y^{\frac{v}{2}-1} \exp\left(-\frac{y}{2}\right)$$

which is a chi square distribution with v degrees of freedom. Since $y = 2\gamma x$,

$$\sqrt{2\gamma x} = \sqrt{\chi_v^2} \sim N\left(\frac{\sqrt{2}}{2}\sqrt{2v-1}, \frac{1}{2}\right)$$

Recall again that $x \sim N(\mu, \sigma^2) \Rightarrow cx \sim N(c\mu, c\sigma^2)$ to get a more convenient expression.

$$c = \frac{1}{\sqrt{2\gamma}} \Rightarrow \sqrt{x} \sim N\left(\frac{1}{2}\sqrt{\frac{1}{\gamma}(2\nu - 1)}, \frac{1}{4\gamma}\right)$$

$$\nu = 2\alpha \Rightarrow \sqrt{x} \sim N\left(\frac{1}{2}\sqrt{\frac{1}{\gamma}(4\alpha - 1)}, \frac{1}{4\gamma}\right)$$

In summary, if x follows the gamma distribution with parameters α and γ, its square root will follow *approximately* a normal distribution with the indicated mean and variance. Application of this model to actual particle data (Levinson et al. 2001) shows an excellent fit of the transformed data to a normal distribution.

There is, however, a significant difference between the one-sided process performance indices from the gamma distribution (1.426) and normal distribution of the transformed data (1.656). The implied nonconforming fractions are respectively:

$\Phi(-3 \times 1.426) = 9.44 \times 10^{-6}$ (9.44 defects per million opportunities [DPMO])

$\Phi(-3 \times 1.656) = 3.39 \times 10^{-7}$ (0.339 DPMO)

The estimate from the transformed data is therefore off by more than an order of magnitude despite the apparent goodness of fit. If the quantiles of the transformed data are not identical to those of the raw data under the original distribution (in this case, the gamma distribution), the transformation is not suitable for process capability calculations.

As another example, consider 100 measurements from a simulated gamma distribution with $\alpha = 8$ and $\gamma = 2$ (Gamma_a8_g2 in the simulated data spreadsheet), and assume that the upper specification is 10 (e.g., parts per million of an impurity or trace contaminant). As will be shown below, the best fit parameters of the gamma distribution are $\alpha = 7.8434$ and $\gamma = 1.8578$, and there are 1707 defects per million opportunities above the upper specification limit.

As shown by Figure 2.1A, the square roots of the measurements conform quite well to a normal distribution with a mean of 2.022 and a standard deviation of 0.3673 (variance 0.1349), although the fact that a series of points dips below the regression line near the center shows that the fit is not perfect. Recalling that

$$\sqrt{x} \sim N\left(\frac{1}{2}\sqrt{\frac{1}{\gamma}(4\alpha - 1)}, \frac{1}{4\gamma}\right),$$

the mean and variance based on $\alpha = 7.8434$ and $\gamma = 1.8578$ should be 2.022 and 0.1346, respectively, and this is true except for the last significant figure of the variance.

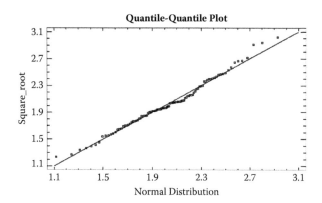

FIGURE 2.1A
Q-Q plot, square root of measurements from gamma distribution.

The calculated nonconforming fraction is, however,

$$1 - \Phi\left(\frac{\sqrt{10} - 2.022}{.3673}\right)$$

or 1222 nonconformances per million opportunities versus 1707 from the underlying gamma distribution. The reason appears to be that the relative difference between the cumulative densities from the gamma distribution and its transformation increases as one gets further into the extreme tail area, as shown in Figure 2.1B. In other words, *the transformation holds up very well in the region of interest for statistical process control (SPC) purposes, but it begins to break down in the region of interest for the nonconforming fraction.*

This underscores the lesson that, even if transformed data conform sufficiently to a normal distribution for use on a traditional control chart, the transformation will not reliably predict the nonconforming fraction and therefore the process performance index. It is therefore necessary to fit the underlying distribution to the data, and the rest of this chapter will apply the following procedure.

General Procedure for Nonnormal Distributions

The following series of steps applies to statistical process control and capability analysis for all nonnormal distributions.

1. Identify the best distribution to model the data.
   - The application will often suggest an appropriate distribution. Undesirable random arrivals such as particles or impurities suggest, for example, that the gamma distribution will apply.

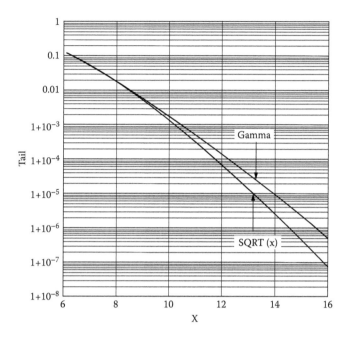

FIGURE 2.1B
Tail areas: Gamma distribution and square root transformation.

- Minitab and StatGraphics have built-in procedures to compare several distributions.

2. Fit a candidate distribution to the data.

3. Perform goodness-of-fit tests.

- This book recommends visual techniques, such as like the histogram and quantile-quantile plot, and supporting quantitative techniques, such as the chi square test.
- Remember that a goodness-of-fit test can never prove that a distribution is appropriate; it can only prove that it is not. Acceptable test results in conjunction with a physical explanation (such as undesirable random arrivals) for the distribution in question do, however, suggest that the selection is a good one.

4. Set up a control chart that uses appropriate quantiles of the fitted distribution.

- Remember that a prerequisite of the distribution fitting process is that the underlying process be in control. The database should not contain measurements with assignable cause variation, and these will often make their presence known on the quantile-quantile plot and histogram as well as on the control chart.

5. Use the distribution's parameters to calculate nonconforming fractions and corresponding process performance indices.

The next section will treat the gamma distribution and a procedure with which to fit it to process data.

The Gamma Distribution

The gamma distribution is the continuous analogue to the Poisson distribution. The probability density function for the 3-parameter gamma distribution is

$$f(x) = \frac{\gamma^\alpha}{\Gamma(\alpha)}(x-\delta)^{\alpha-1}\exp(-\gamma(x-\delta)) \qquad (2.2)$$

where α is the shape parameter, γ is the scale parameter, and δ is the threshold parameter. The latter corresponds to the *guarantee time* in reliability statistics, but in the case of a gamma distribution, the context is far more frequently the minimum number of particles, impurities, or defects that the process can produce. Lawless (1982) describes how to use the maximum likelihood method to optimize the parameters for this distribution and others.

Maximum Likelihood Distribution Fitting

Given any probability density function $f(X,\theta)$ where X is the data vector $x_1, x_2, \ldots, x_n$ and θ is the vector of parameters (such as μ and σ for the normal distribution, or α, γ, and δ for the gamma distribution), the *likelihood function* is:[1]

$$L(\theta) = \prod_{i=1}^{n} f(x_i|\theta) \quad \text{and} \quad \ln L(\theta) = \sum_{i=1}^{n} \ln(f(x_i|\theta)) \qquad \text{(Set 2.3)}$$

The log likelihood function is generally superior for computational purposes because there is less chance of a computational overrun. The maximum likelihood approach seeks the set of parameters that maximizes $L(\theta)$ or $\ln L(\theta)$.

Consider, for example, the familiar normal distribution:

$$f(x) = \frac{1}{\sqrt{2\pi}\sigma}\exp\left(-\frac{1}{2\sigma^2}(x-\mu)^2\right)$$

$$L(\mu,\sigma) = \prod_{i=1}^{n} \frac{1}{\sqrt{2\pi}\sigma}\exp\left(-\frac{1}{2\sigma^2}(x_i-\mu)^2\right)$$

This is a situation in which the natural log of the likelihood function is far more convenient.

$$\ln L(\mu,\sigma) = \sum_{i=1}^{n} \ln\left(\frac{1}{\sqrt{2\pi}\sigma}\right) - \frac{1}{2\sigma^2}(x_i - \mu)^2$$

$$= n\ln\left(\frac{1}{\sqrt{2\pi}\sigma}\right) - \sum_{i=1}^{n} \frac{1}{2\sigma^2}(x_i - \mu)^2$$

$$= n\ln\left(\frac{1}{\sqrt{2\pi}}\right) - n\ln\sigma - \sum_{i=1}^{n} \frac{1}{2\sigma^2}(x_i - \mu)^2$$

$$\frac{\partial \ln L(\mu,\sigma)}{\partial \mu} = \sum_{i=1}^{n} \frac{1}{\sigma^2}(x_i - \mu)$$

Set the partial derivative equal to zero to maximize this function for μ.

$$\sum_{i=1}^{n} \frac{1}{\sigma^2}(x_i - \mu) = 0 \Rightarrow \sum_{i=1}^{n} x_i - n\mu = 0$$

$$\mu = \frac{1}{n}\sum_{i=1}^{n} x_i$$

The estimate for the mean is simply the sum of the individual measurements divided by the number of measurements, as is expected. Guttman, Wilkes, and Hunter (1982, 176) add that the maximum likelihood estimator for the variance is not the familiar unbiased estimator with $(n-1)$ in the denominator. Its derivation is as follows:

$$\ln L(\mu,\sigma) = n\ln\left(\frac{1}{\sqrt{2\pi}}\right) - n\ln\sigma - \sum_{i=1}^{n} \frac{1}{2\sigma^2}(x_i - \mu)^2$$

$$= n\ln\left(\frac{1}{\sqrt{2\pi}}\right) - \frac{n}{2}\ln\sigma^2 - \sum_{i=1}^{n} \frac{1}{2\sigma^2}(x_i - \mu)^2$$

$$\frac{\partial \ln L(\mu,\sigma)}{\partial \sigma^2} = -\frac{n}{2\sigma^2} + \sum_{i=1}^{n} \frac{1}{2(\sigma^2)^2}(x_i - \mu)^2$$

$$0 = -\frac{n}{2\sigma^2} + \sum_{i=1}^{n} \frac{1}{2(\sigma^2)^2}(x_i - \mu)^2 \Rightarrow$$

$$n = \frac{1}{\sigma^2}\sum_{i=1}^{n}(x_i - \mu)^2 \Rightarrow \sigma^2 = \frac{1}{n}\sum_{i=1}^{n}(x_i - \mu)^2$$

The denominator of n assumes that the parameters are for the entire population, while a denominator of $(n-1)$ is the unbiased estimator for a sample. The primary purpose of this exercise is, however, to show how the maximum likelihood method works. The reference cites a similar exercise with the Bernoulli distribution that yields the number of nonconforming parts divided by the total number of parts as the maximum likelihood estimate of the binomial distribution's parameter. This again is the expected outcome.

In the case of the gamma distribution, Lawless (1982, 204–206) gives the following procedure for maximum likelihood estimators. The first step is to find the arithmetic and geometric averages of the data. The second equation for the geometric mean should avoid any computational overflow that might occur in the first version.

$$\bar{x} = \frac{1}{n}\sum_{i=1}^{n} x_i \quad \text{and} \quad \tilde{x} = \left(\prod_{i=1}^{n} x_i\right)^{\frac{1}{n}} \quad \text{or} \quad \tilde{x} = \exp\left(\frac{1}{n}\sum_{i=1}^{n} \ln(x_i)\right) \qquad \text{(Set 2.4)}$$

Then solve the following equations:

$$\frac{1}{\gamma} = \frac{\bar{x}}{\alpha} \quad \text{and} \quad \ln(\alpha) - \psi(\alpha) = \ln\frac{\bar{x}}{\tilde{x}} \qquad \text{(Set 2.5)}$$

where $\psi(\alpha)$ is the digamma function.

$$\psi(\alpha) = \frac{d\ln\Gamma(\alpha)}{d\alpha} \approx \ln\alpha - \frac{1}{2\alpha} - \frac{1}{12\alpha^2} + \frac{1}{120\alpha^4} - \frac{1}{256\alpha^6} + \cdots$$

This does not, however, solve the problem of what to do about the threshold parameter. Successful results have been obtained from plots of $L(\alpha,\gamma,\delta)$ (or its natural logarithm) versus δ, and from computer search methods that find δ to maximize this function. Note that shape and scale parameters must be recalculated for each threshold parameter, but this is not a real problem for modern high-speed computers. The current versions of Minitab and StatGraphics can fit the three-parameter gamma distribution on their own, but the next section will discuss an algorithm for solving these equations if the user must program a computer to do so. The book also includes a Visual Basic for Applications function that will work in Microsoft Excel. The next section addresses the simpler case of the two-parameter gamma distribution, and the procedure becomes a subroutine for the three-parameter case.

Bisection Algorithm

There are several ways to find the root of an equation, and the Newton-Raphson method is among the most popular. It begins with an initial guess

x_0 to solve the equation $f(x) = 0$, and it iterates the following expression until x ceases to change significantly.

$$x_{i+1} = x_i - \frac{f(x_i)}{f'(x_i)} \quad \text{Newton-Raphson procedure}$$

This procedure essentially draws a tangent to the graph of the function at x_i and traces it back to the axis to get x_{i+1}, and convergence to a solution is often very rapid. If the initial guess is poor, though, it will sometimes not find a solution. Although bisection is not as rapid, it will *always* find the root if (1) it is within the range $[a,b]$ that the user sets initially and (2) there is exactly one x that meets the requirement $f(x) = 0$. The relative computational inefficiency may have been an issue several decades ago but it is not a problem for today's high-speed processors.

The maximum likelihood fit for the gamma distribution requires solution of the equation set

$$\frac{1}{\gamma} = \frac{\overline{x}}{\alpha} \quad \text{and} \quad \ln(\alpha) - \psi(\alpha) = \ln\frac{\overline{x}}{\tilde{x}}$$

Define $f(\alpha) = \ln(\alpha) - \psi(\alpha) - \ln\frac{\overline{x}}{\tilde{x}}$ and also $[\alpha_L, \alpha_R] = [a,b]$ where the subscripts L and R mean "left" and "right." The only way in which this procedure can go wrong is if the solution is outside the user-defined initial range $[a,b]$, and it is easy to test for this before the procedure begins. A good initial estimate for α can be obtained as follows, and the maximum should be somewhat greater to make sure the answer is in fact within the range.

$$\mu_{gamma} = \frac{\alpha}{\gamma} \quad \sigma^2_{gamma} = \frac{\alpha}{\gamma^2} \quad \Rightarrow \quad \frac{\mu^2_{gamma}}{\sigma^2_{gamma}} = \alpha \quad (2.6)$$

Division of the square of the data average by its variance should therefore give a rough idea as to the location of the shape parameter. Then bisect the interval $[a,b]$ as follows (Hornbeck, 1975, 66):

1. If $f(x_L) \times f(x_R) < 0$ then go to step 3, otherwise go to step 2
   - If the functions of x_L and x_R have opposite signs, the root is between x_L and x_R. If not, the root is outside this interval.
   - This condition will always be true in the first iteration unless the root is not in the interval $[a,b]$, so the *TEMP* variable shown below will always be defined. It is useful to perform this test up front to make sure the answer is in fact somewhere between a and b.

2. The root is outside the interval $[x_L, x_R]$: Let $x_R = x_L$, $x_L = TEMP$, and then

$$x_L = \frac{x_L + x_R}{2}.$$

Return to step 1.

3. The root is inside the interval $[x_L, x_R]$: If $x_R - x_L < 2\varepsilon$ where ε is the desired tolerance for the result, then the solution is

$$\frac{x_L + x_R}{2}$$

(stopping rule). Otherwise let $TEMP = x_L$, and then

$$x_L = \frac{x_L + x_R}{2}$$

and then return to step 1.

The following example applies this method to a simple gamma distribution.

Example: Two-Parameter Gamma Distribution

Given 100 random data from a gamma distribution with $\alpha = 8$, $\gamma = 2$, and upper specification limit 10 (Gamma_a8_g2 in the simulated data spreadsheet), fit the distribution and calculate the process performance index. Figure 2.2 illustrates the bisection method in MathCAD. The first part calculates necessary information such as the number of points, their arithmetic and geometric means, and their variance. It also defines the digamma function and makes sure the solution is in the range $[a,b]$ to be bisected.

Since $f(a) \times f(b)$ is negative, it means there is an α in the range $[a,b]$ that will yield a result of zero. The second part graphs $f(a)$ as a function of a, and this shows that the solution will be somewhat greater than 7.5. The program loop performs the bisection; note that the tolerance is built into the loop.

Figure 2.3, a contour plot of $\sum_{i=1}^{n} \ln(f_{gamma}(x_i | \alpha, \gamma))$ versus α (x axis) and γ (y axis),[2] shows how maximum likelihood optimization works for this dataset, and illustrates how maximum likelihood estimation works in general. The contour lines correspond to the log likelihood function as a function of the shape and scale parameters, and the maximum is slightly greater than −178.6. The set of parameters that go with this maximum result in the best fit for the dataset, and bisection is but one of several ways to locate this point.

To find the process performance index, compute the tail area above 10. This can be programmed in MathCAD, or calculated by Excel's GAMMADIST function as

Initial statistics

$$\begin{bmatrix} n \\ x\_bar \\ x\_geom \end{bmatrix} := \begin{bmatrix} rows(x) \\ mean(x) \\ \exp\left(\dfrac{1}{rows(x)} \cdot \displaystyle\sum_{i=1}^{rows(x)} \ln(x_i)\right) \end{bmatrix} \qquad \begin{bmatrix} n \\ x\_bar \\ x\_geom \end{bmatrix} = \begin{bmatrix} 100 \\ 4.222 \\ 3.956 \end{bmatrix}$$

Variance of the data set and estimate for alpha

$$var := \frac{1}{n-1} \cdot \sum_{i=1}^{n} (x_i - x\_bar)^2 \quad alpha_0 := \frac{x\_bar^2}{var} \quad alpha_0 = 7.515$$

Equation to be solved for alpha

Digamma function $\qquad \psi(y) := \dfrac{d}{dy} \ln(\Gamma(y))$

$$f(\alpha) := \ln(\alpha) - \psi(\alpha) - \ln\left(\frac{x\_bar}{x\_geom}\right)$$

Range to bisect $\begin{bmatrix} a \\ b \end{bmatrix} := \begin{bmatrix} 5 \\ 10 \end{bmatrix} \qquad f(a) \cdot f(b) = -5.453 \cdot 10^{-4}$

FIGURE 2.2A
Setup for bisection fit of gamma distribution parameters.

follows: =1–GAMMADIST(10,7.8434,1/1.8578,1), noting that Excel uses the reciprocal of gamma as the scale parameter (as does Minitab). The result is 0.0017070 (1707 defects per million opportunities). The process performance index is the absolute value of the corresponding standard normal deviate divided by 3, for example, =ABS(NORMSINV(0.0017070)/3) in Excel, or 0.976.

For comparison, treat the data as a normal distribution. The variance of the 100 data points is 2.372, so the standard deviation is 1.540, and the grand average is 4.222.

$$PPU_{normal} = \frac{10 - 4.222}{3 \times 1.540} = 1.251$$

This promises a nonconforming fraction of 88.4 defects or nonconformances per million opportunities (NORMSDIST(–3*1.25) in Excel) versus the actual 1707; in other words, *assumption of normality yields an expected nonconforming fraction that is inaccurate by more than an order of magnitude.*

Minitab 15 yields $\alpha = 7.84337$ and $\beta = 0.53827$, where β is the inverse of γ. Figure 2.4 shows its capability analysis for an upper specification limit of 10, with $PPU = 0.97$. The distribution fitting routine also provides a quantile-quantile plot.

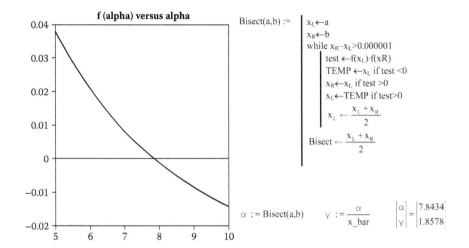

FIGURE 2.2B
Bisection fit of gamma distribution parameters.

StatGraphics Centurion yields $\alpha = 7.84337$ and $\gamma = 1.85782$. It also performs the chi square goodness-of-fit test as shown in Table 2.1 (given a histogram that begins with 10 cells). The chi square table has seven cells, and because two parameters were estimated from the data, the test statistic has four degrees of freedom ($7 - 2 - 1 = 4$).

StatGraphics Centurion also produces the capability analysis in Table 2.2 if directed to use the gamma distribution.

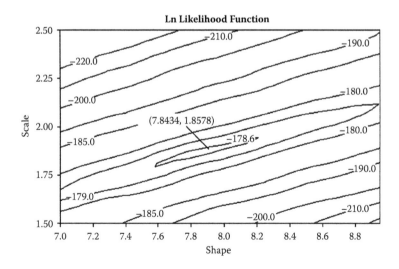

FIGURE 2.3
Maximum likelihood optimization of a gamma distribution.

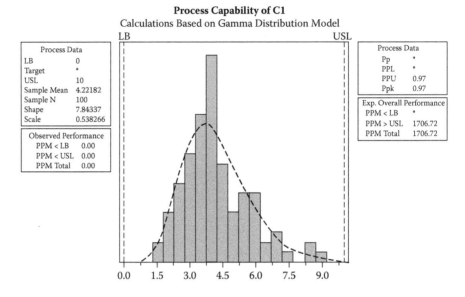

FIGURE 2.4
Minitab capability analysis for two-parameter gamma distribution.

StatGraphics finds the same nonconforming fraction (1707 defects per million opportunities) and long-term process performance index. The analysis also displays a quantile-quantile plot, and the chi square test for goodness of fit can also be performed.

It has already been shown that the square root transformation does not yield an accurate estimate of the nonconforming fraction. The Johnson transformation (Chou, Polansky, and Mason, 1998) is a family of distributions that transform data to the standard normal distribution, and Table 2.3 shows the available transformations.[3]

TABLE 2.1

Chi Square Goodness-of-Fit Test, Two-Parameter Gamma Data

	Lower Limit	Upper Limit	Observed Frequency	Expected Frequency	Chi-Squared
At or below		2.0	5	4.13	0.18
	2.0	3.0	16	17.56	0.14
	3.0	4.0	31	27.12	0.55
	4.0	5.0	22	23.94	0.16
	5.0	6.0	13	14.98	0.26
	6.0	7.0	6	7.43	0.28
Above	7.0		7	4.84	0.96

Note: Chi squared = 2.52919 with 4 d.f. P-Value = 0.639416

TABLE 2.2

StatGraphics Capability Analysis, Two-Parameter Gamma Data
USL = 10.0
LSL = 0.0

	Short-Term Capability	Long-Term Performance
Sigma (after normalization)	0.982961	1.0
C_{pk}/P_{pk} (upper)	0.99286	0.975943
DPM	1448.0	1706.72

Note: Based on 6.0 sigma limits in the normalized metric. Short-term sigma estimated from average moving range.

Minitab 15 fits the unbounded S_U transformation to this dataset and obtains

$$z(x) = -1.51574 + 1.89910 \times \sinh^{-1}\left(\frac{x - 2.27278}{1.94699}\right)$$

along with a relatively promising normal probability plot (Figure 2.5). Then, if the upper specification limit is 10, $z(10) = 2.449$ and the corresponding non-conforming fraction is 0.007166 (7166 DPMO), or more than four times too great. This underscores the suspicion that, while transformed data may perform better than raw data under the normality assumption, these transformations do not provide reliable process capability or process performance estimates.

As a final note, it is possible to fit the two-parameter gamma distribution in Microsoft Excel, which is more widely available than MathCAD, Minitab, or StatGraphics. It can be programmed as shown here, or the user can employ a Visual Basic for Applications program to do the job automatically.

TABLE 2.3

Johnson Transformations

	Transformation	
Bounded S_B	$z = \gamma + \eta \ln\left(\dfrac{x - \varepsilon}{\lambda + \varepsilon - x}\right)$	Requires x to be between ε and $\varepsilon + \lambda$, noting the argument of the natural log function. The other arguments are fitted parameters.
Lognormal S_L	$z = \gamma + \eta \ln(x - \varepsilon)$	Requires $x > \varepsilon$, again noting the argument of the natural log
Unbounded S_U	$z = \gamma + \eta \sinh^{-1}\left(\dfrac{x - \varepsilon}{\lambda}\right)$	$\sinh^{-1} =$ inverse hyperbolic sine

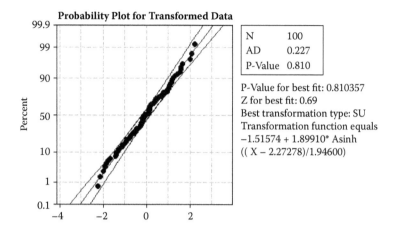

FIGURE 2.5
Johnson transformation and probability plot.

Two-Parameter Gamma Distribution Fit, Microsoft Excel

There are at least three ways to fit the gamma distribution in Microsoft Excel:

1. Use cell formulas and the Solver or Goal Seek routine to find the shape parameter, after which the scale parameter may be found directly.
2. Use Visual Basic for Applications to perform all the necessary calculations.
3. The section on left-censored Weibull and gamma distributions shows that a response surface analysis method can optimize two parameters simultaneously provided that the response surface has a unique maximum and no local maximums at which the algorithm might settle. Figure 2.3 shows that this is the case.

The first approach begins with calculation of the arithmetic and geometric means of the data. Given 100 data (the same set used above) in cells B6 through B105:

$$\text{Arithmetic mean} = \text{SUM(B6:B105)}/100$$

$$\text{Geometric mean} = \{\text{EXP(SUM(LN(B6:B105)/100))}\}$$

where the brackets indicate array summation and are obtained by pressing **Ctrl-Shift-Enter**. Let the arithmetic and geometric means be stored in cells F5 and F6, respectively. The difficult part comes from the fact that Excel has no built-in digamma function, but there is a workaround for this. Place the following expression in cell G7, while cell F7 will hold the alpha parameter.

=LN(F7)−(LN(1/(GAMMADIST(1,F7+0.000001,1,0)*EXP(1)))−LN(1/(GAMMA DIST(1,F7,1,0)*EXP(1))))/0.000001−LN(F5/F6)

GAMMADIST$(x,\alpha,\beta,0)$ returns

$$\frac{1}{\beta^{\alpha}\Gamma(\alpha)}x^{\alpha-1}\exp\left(-\frac{x}{\beta}\right)$$

noting that β is the reciprocal of α. Therefore,

$$\text{GAMMADIST}(1,F7,1,0)\times\text{EXP}(1)=\frac{1}{1\times\Gamma(F7)}1^{F7-1}\exp(-1)\exp(1)=\frac{1}{\Gamma(F7)}$$

and its reciprocal is then $\Gamma(F7)$. Finally,

$$\frac{\ln(\Gamma(F7+0.000001))-\ln(\Gamma(F7))}{0.000001}\approx\frac{d\ln(\alpha)}{d\alpha}=\psi(\alpha)$$

evaluated for the α in cell F7.

Then use Excel's SOLVER tool as shown in Figure 2.6 to find α to make the expression in cell G7 equal zero; the result of 7.8434 matches those from the MathCAD bisection routine, Minitab, and StatGraphics. GOAL SEEK also works and under circumstances in which SOLVER fails to converge on a solution. In other words, if SOLVER won't work, try GOAL SEEK and, if GOAL SEEK can't find a solution, SOLVER might.

The Visual Basic for Applications function FitGamma fits the two-parameter gamma distribution to a selected range of cells. The user can copy and paste the module from the Notepad files on the user disk or import the .bas file by using **Tools > Macro > Visual Basic Editor**. A professional programmer could doubtlessly improve on this routine, but we have verified that it can read data from multiple columns, and even columns and rows with blank cells. For example, FitGamma(H6:J43,20) locates the 100 data in the cell range H6 through J43, even with columns of unequal size and a row with measurements in only two columns, and returns $\alpha = 7.8434$, $\gamma = 1.8578$; the same results that

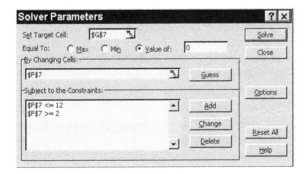

FIGURE 2.6
Use of Microsoft Excel SOLVER to fit two-parameter gamma distribution.

MathCAD, Minitab, and StatGraphics returned for this dataset. The user must provide a maximum value of the shape parameter for the bisection routine, and the program will report if the solution is outside the bisection range.

The three-parameter case is somewhat more complicated, but the current versions of Minitab and StatGraphics have built-in routines with which to handle it. The following section will illustrate bisection routines in MathCAD that are adaptable to other programming languages if necessary.

Three-Parameter Gamma Distribution

The threshold parameter of the three-parameter gamma distribution must be less than the smallest value in the dataset. The challenge is to select the one that maximizes the likelihood function or its natural log:

$$\ln L(\alpha, \gamma, \delta) = \sum_{i=1}^{n} \ln\left(\frac{\gamma^{\alpha}}{\Gamma(\alpha)}(x_i - \delta)^{\alpha-1} \exp(-\gamma(x_i - \delta)) \right)$$

$$= \alpha \times n \times \ln(\gamma) - n \times \ln(\Gamma(\alpha)) + \sum_{i=1}^{n}(\alpha - 1)\ln(x_i - \delta) - \gamma(x_i - \delta)$$

This requires maximization of the likelihood function with respect to the threshold parameter, which suggests solution of

$$\frac{d \ln L(\alpha, \gamma, \delta)}{d\delta} \approx \frac{\ln L(\alpha, \gamma, \delta + \Delta) - \ln L(\alpha, \gamma, \delta)}{\Delta} = 0$$

where Δ is a very small interval. A bisection routine for the entire process appears in Figure 2.7. The dataset is the same as for the previous example; even though it was simulated without a threshold parameter, application of one yields a very good fit.

A plot of the log likelihood versus threshold parameter shows that the optimum value will be somewhere near 0.6. Figure 2.7C shows *visually* what the maximum likelihood algorithm seeks to achieve, while Figure 2.7D shows the mathematical optimization.

This shows that the optimum threshold parameter is about 0.63. The final step is to compute and report all the parameters, as shown in Figure 2.7E.

The result is a gamma distribution with shape parameter $\alpha = 5.482$, scale parameter $\gamma = 1.526$, and threshold parameter $\delta = 0.6295$. The practical implication is that the process cannot deliver a product with less than, for example, 0.6295 parts per million of an impurity if ppm is the unit of measurement.

Minitab 15 and, as shown by Table 2.4, StatGraphics Centurion yield almost identical results.

Figure 2.8 shows additional results from StatGraphics *that we would expect to see in a process capability report from a vendor or characterization of a*

Initial statistics

$$\begin{bmatrix} n \\ \delta_{min} \\ \delta_{max} \end{bmatrix} := \begin{bmatrix} \text{rows}(x) \\ 0 \\ \min(x) \cdot 0.9999 \end{bmatrix} \qquad \begin{bmatrix} n \\ \delta_{min} \\ \delta_{max} \end{bmatrix} = \begin{bmatrix} 100 \\ 0 \\ 1.53 \end{bmatrix}$$

The maximum allowable threshold parameter must be slightly less

than the minimum observation in the data set because a calculation

involves the natural log of x minus the threshold parameter.

Preliminary graph of the log likelihood function versus threshold parameter

$$\text{f\_gamma}(x,\alpha,\gamma,\delta) := \frac{\gamma^a}{\Gamma(\alpha)} \cdot (x-\delta)^{a-1} \cdot \exp(-\gamma \cdot (x-\delta))$$

Digamma function $\qquad \psi(y) := \dfrac{d}{dy} \ln(\Gamma(y))$

Equation to be solved

for the shape parameter $\quad f(\alpha, \text{avg}, \text{geom\_mean}) := \ln(\alpha) - \psi(\alpha) - \ln\left(\dfrac{\text{avg}}{\text{geom\_mean}}\right)$

FIGURE 2.7A
Threshold parameter optimization (MathCAD).

process in a factory. Table 2.5 is the chi square test for goodness of fit from StatGraphics, and Table 2.6 is from the Visual Basic for Applications function =ChiSquare_Gamma(D7:F44,10,5.482,1.526,0.6295,"C:\TEMP\ChiSquare.txt") where the arguments are the data range, number of cells, alpha, gamma, delta, and the location of the output file. It is almost identical to the one for the normal distribution except it uses Excel's GAMMADIST function to calculate the expected counts in each cell. Remember that this function uses the reciprocal of gamma as its scale parameter; to program it manually, use GAMMADIST(x, alpha, 1/gamma, 1) where x is the cell floor or ceiling. The last argument (1) tells the spreadsheet to use the cumulative distribution as opposed to the probability density function.

Note that, as three parameters were estimated from the data, there are $8 - 4 = 4$ degrees of freedom.[4] The test statistic is sufficiently low to pass the test easily. Combination of cells in Table 2.6 to make the expected frequencies 5 or more, yields equally acceptable results as shown in Table 2.5B.

Combination of the last three cells in Table 2.6 to meet the requirement that the expected cell count be no less than 5 yields a test statistic of 1.784 for eight cells. Excel's built-in Chart Wizard can then display a bar histogram of the observed counts with lines for the expected counts.

The Visual Basic for Applications (VBA) function QQ_Plot works as follows: QQ_Plot(D7:F44,"Gamma",5.482,1.526,0.6295,"C:\TEMP\QQ_Gamma. txt") returns a slope of 1.004, intercept −0.01465, and correlation 0.9961

$$
\text{1nL}(d) := \begin{vmatrix}
\text{x\_bar} \leftarrow \text{mean}(x) \\[4pt]
\text{variance} \leftarrow \dfrac{1}{n-1} \cdot \sum_{i=1}^{n} (x_i - \text{x\_bar})^2 \\[4pt]
\text{x\_bar} \leftarrow \text{x\_bar} - d \\[4pt]
\text{x\_geom} \leftarrow \exp\left[\left(\dfrac{1}{n} \right) \cdot \sum_{i=1}^{n} \ln\left(x_i - d\right) \right] \\[4pt]
\alpha_0 \leftarrow \dfrac{\text{x\_bar}^2}{\text{variance}} \\[4pt]
x_L \leftarrow \dfrac{\alpha_0}{5} \\[4pt]
x_R \leftarrow 5 \cdot \alpha_0 \\[4pt]
\text{while } x_R - x_L > 0.000001 \\[4pt]
\quad \begin{vmatrix}
\text{test} \leftarrow f(x_L, \text{x\_bar}, \text{x\_geom}) \cdot f(x_R, \text{x\_bar}, \text{x\_geom}) \\[4pt]
\text{TEMP} \leftarrow x_L \text{ if test} < 0 \\[4pt]
x_R \leftarrow x_L \text{ if test} > 0 \\[4pt]
x_L \leftarrow \text{TEMP if test} > 0 \\[4pt]
x_L \leftarrow \dfrac{x_L + x_R}{2}
\end{vmatrix} \\[4pt]
\alpha \leftarrow \dfrac{x_L + x_R}{2} \\[4pt]
\gamma \leftarrow \dfrac{\alpha}{\text{x\_bar}} \\[4pt]
\text{1nL} \leftarrow \sum_{i=1}^{n} \ln\left(\text{f\_gamma}\left(x_i, \alpha, \gamma, d\right) \right)
\end{vmatrix}
$$

Note that the arithmentic and geomentric averages will depend on the threshold parameter in the function argument. Use x_bar to get the variance, and then deduct the threshold parameter as shown.

Bisection limits of 1/5 to 5 times the estimated alpha value should ensure convergance.

This part of the algorithm finds the shape and scale parameters that go with the threshold parameter in the function argument.

FIGURE 2.7B
Log likelihood as a function of threshold parameter.

along with 3 columns of data that result in a graph similar to the one from StatGraphics when plotted as a scatter chart.

Three-Parameter Gamma Distribution Fit, Microsoft Excel

It is also possible, albeit somewhat labor intensive, to fit the three-parameter gamma distribution on an Excel spreadsheet as shown in Table 2.7. The basic idea is to bisect the range in which the threshold parameter can exist until the

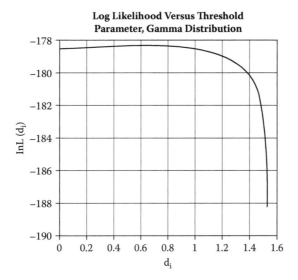

FIGURE 2.7C
Log likelihood versus threshold parameter.

Bisect the interval for the point at which the slope is zero (maximum)

$$\Delta := 0.000001 \quad d\_lnL(d) := \frac{lnL(d+\Delta) - lnL(d)}{\Delta}$$

Numerical estimate for the
first derivative at any point

$d\_ML(a,b) := \begin{vmatrix} x_L \leftarrow a \\ x_R \leftarrow b \\ \text{while } x_R - x_L > 0.000001 \\ \quad \begin{vmatrix} \text{test} \leftarrow d\_lnL(x_L) \cdot d\_lnL(x_R) \\ \text{TEMP} \leftarrow x_L \text{ if test} < 0 \\ x_R \leftarrow x_L \text{ if test} > 0 \\ x_L \leftarrow \text{TEMP if test} > 0 \\ x_L \leftarrow \dfrac{x_L + x_R}{2} \end{vmatrix} \\ d\_ML \leftarrow \dfrac{x_L + x_R}{2} \end{vmatrix}$

The objective is to find the
threshold parameter at
which the first derivative
is zero.

$\delta := d\_ML(\delta_{min}, \delta_{max})$

$\delta = 0.63$

FIGURE 2.7D
Bisection for optimum threshold parameter.

Now report the shape and scale parameters for this threshold parameter

$$x\_bar := \text{mean}(x) \qquad x\_geom := \exp\left[\left(\frac{1}{n}\right) \cdot \sum_{i=1}^{n} \ln(x_i - \delta)\right]$$

$$\text{variance} := \frac{1}{n-1} \cdot \sum_{i=1}^{n}\left[(x_i - \delta) - x\_bar\right]^2 \qquad x\_bar := x\_bar - \delta \qquad \text{Again, deduct the threshold from the average after computation of the variance.}$$

$$\alpha_0 := \frac{x\_bar^2}{\text{variance}} \qquad f(\alpha) := \ln(\alpha) - \psi(\alpha) - \ln\left(\frac{x\_bar}{x\_geom}\right) \qquad \text{The bisection interval will range from 1/5 of the estimated shape parameter to five times the estimated shape parameter.}$$

$$\text{Bisect}(a,b) := \begin{vmatrix} x_L \leftarrow a \\ x_R \leftarrow b \\ \text{while } x_R - x_L > 0.000001 \\ \quad \begin{vmatrix} \text{test} \leftarrow f(x_L) \cdot f(x_R) \\ \text{TEMP} \leftarrow x_L \text{ if test} < 0 \\ x_R \leftarrow x_L \text{ if test} > 0 \\ x_L \leftarrow \text{TEMP if test} > 0 \\ x_L \leftarrow \frac{x_L + x_R}{2} \end{vmatrix} \\ \text{Bisect} \leftarrow \frac{x_L + x_R}{2} \end{vmatrix}$$

$$\alpha := \text{Bisect}\left(\frac{\alpha_0}{5}, 5 \cdot \alpha_0\right)$$

$$\gamma := \frac{\alpha}{x\_bar}$$

$$\begin{bmatrix} \alpha \\ \gamma \\ \delta \end{bmatrix} = \begin{bmatrix} 5.48224 \\ 1.52612 \\ 0.62954 \end{bmatrix}$$

FIGURE 2.7E
Display all parameters for the gamma distribution.

TABLE 2.4

StatGraphics Results,
Three-Parameter Fit
100 values ranging
from 1.53048 to 9.11314

Fitted Distributions

Gamma (3-Parameter)
shape = 5.48223
scale = 1.52612
lower threshold = 0.629542

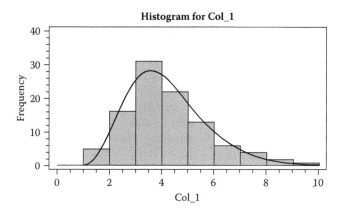

FIGURE 2.8A
Histogram, three-parameter gamma distribution.

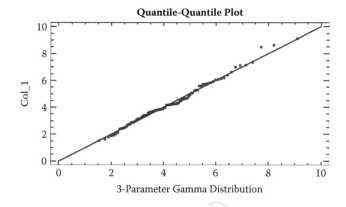

FIGURE 2.8B
Quantile-Quantile plot, three-parameter gamma distribution.

TABLE 2.5A

Chi Square Goodness-of-Fit Test, Three-Parameter Gamma Distribution

	Lower Limit	Upper Limit	Observed Frequency	Expected Frequency	Chi-Squared
At or below		2.0	5	3.65	0.50
	2.0	3.0	16	18.63	0.37
	3.0	4.0	31	27.57	0.43
	4.0	5.0	22	23.24	0.07
	5.0	6.0	13	14.35	0.13
	6.0	7.0	6	7.29	0.23
	7.0	8.0	4	3.24	0.18
Above	8.0		3	2.05	0.45

Note: Chi Squared = 2.34847 with 4 d.f. P-Value = 0.671959

likelihood function or its natural log is maximized. The results of $\alpha = 5.490$, $\gamma = 1.526$, and $\delta = 0.63$ are very close to the exact solutions shown above.

The Visual Basic for Applications function FitGamma3 reproduces the results shown in Table 2.7, and it is included as FitGamma3.bas. The arguments are the cell selection of the data to which the data are to be fit and the maximum shape parameter. The function itself determines the maximum allowable threshold parameter from the data.

Once a gamma distribution has been fit to the data, the next step is to construct a control chart.

Control Chart for the Gamma Distribution

The control chart should behave just like a Shewhart chart in terms of false alarm risks, although the chart's power (e.g., Average Run Length) will generally differ from that of a Shewhart chart for any given process shift. This means that the tail area for each control limit should be 0.00135 while the center line should be the *median* (50th percentile) as opposed to the mean.

TABLE 2.5B

Chi Square Test, Expected Frequencies of 5 or Greater

	Lower Limit	Upper Limit	Observed Frequency	Expected Frequency	Chi-Squared
At or below	2.0	3.0	21	22.28	0.07
	3.0	4.0	31	27.57	0.43
	4.0	5.0	22	23.24	0.07
	5.0	6.0	13	14.35	0.13
	6.0	7.0	6	7.29	0.23
Above	7.0		7	5.29	0.55
					$\chi^2 = 1.48$

TABLE 2.6

Chi Square Test from VBA Function ChiSquare_Gamma
Chi square test for goodness of fit, gamma distribution
VBA.xls
Gamma_a8_g2
100 data 10 Cells

Cell	Floor	Ceiling	Observed	Expected	Chi Square
1	1.530	2.289	7	7.32	0.014
2	2.289	3.047	14	16.169	0.291
3	3.047	3.805	24	21.004	0.427
4	3.805	4.564	22	19.484	0.325
5	4.564	5.322	12	14.669	0.486
6	5.322	6.080	9	9.586	0.036
7	6.080	6.838	5	5.655	0.076
8	6.838	7.597	4	3.088	0.269
9	7.597	8.355	0	1.587	1.587
10	8.355	9.113	3	1.438	1.696
			100	100	5.207

The two are identical for the normal distribution but not for most other distributions.

Consider again a gamma distribution with shape parameter $\alpha = 5.482$, scale parameter $\gamma = 1.526$, and threshold parameter $\delta = 0.6295$. StatGraphics offers a built-in function, Critical Values, in which the user can specify quantiles of the distribution. The necessary information appears in Table 2.8.

This shows that the control limits should be [1.265,10.584] and the center line should be 4.006. Excel's GAMMAINV function returns similar results, for example, =0.6295+GAMMAINV(0.99865,5.482,1/1.526) returns 10.5848. Figure 2.9A shows how StatGraphics creates a control chart for this dataset.

Minitab 15's Process Capability Sixpack yields identical results as shown in Figure 2.9B, although it uses the average of the data as opposed to the distribution's median as the center line.

StatGraphics and Minitab can fit a distribution to the data very quickly, assess the goodness of fit, and display a control chart for the quality practitioner's inspection. For deployment of the control chart to the factory floor, we must, however, remember Heinlein's admonition. The weapon or tool should be as simple as possible, perhaps even as simple as a stone ax, as opposed to something that requires the user to read a vernier. A spreadsheet can meet the requirement of a *visual control*, which makes the status of the process obvious without the need to interpret data or, to use Heinlein's words, read a vernier.

TABLE 2.7

Use of Microsoft Excel to Tit a Three-Parameter Gamma Distribution

	D Delta	E Bisecting	F	G Mean	H Geom	I	J Alpha	K Gamma	L logL	M
6	0.7500	0.0000	1.5000	3.4718	3.1352	–0	5.07257	1.46107	–178.34877	
7	0.3750	0.7500	0.0000	3.8468	3.5506	0	6.32678	1.64468	–178.38422	
8	1.1250	0.7500	1.5000	3.0968	2.7001	–0	3.80878	1.22990	–178.75263	
9	0.5625	0.7500	0.3750	3.6593	3.3445	0	5.66051	1.54688	–178.34032	New best
	0.6563	0.5625	0.7500	3.5656	3.2403	0	5.38150	1.50930	–178.33432	
	0.4688	0.5625	0.3750	3.7531	3.4479	0	6.03856	1.60897	–178.35422	
	0.6094	0.5625	0.6563	3.6124	3.2925	0	5.53704	1.53278	–178.33419	New best
	0.5157	0.5625	0.4688	3.7062	3.3963	0	5.84698	1.57763	–178.34553	
	0.5626	0.6094	0.5157	3.6593	3.3445	0	5.70647	1.55946	–178.33760	
	0.6329	0.6094	0.6563	3.5890	3.2664	–0	5.47060	1.52428	–178.333612	New best
	0.6212	0.6329	0.6094	3.6007	3.2794	0	5.50058	1.52765	–178.333776	
	0.6446	0.6329	0.6563	3.5772	3.2533	0	5.40411	1.51070	–178.334386	
	0.6271	0.6329	0.6212	3.5948	3.2729	0	5.48940	1.52705	–178.333610	New best
	0.6300	0.6271	0.6329	3.5918	3.2696	–0	5.48064	1.52587	–178.333602	New best

D6=(E6+F6)/2

G6=SUM(B$6:B$105)/100–D6, i.e., the average minus the threshold parameter

H6=[EXP(SUM(LN(B$6:B$105–D6)/100))] Geometric mean after deduction of the threshold parameter, array summation

I6=LN(I6)–(LN(1/(GAMMADIST(1,J6+0.000001,1,0)*EXP(1)))–LN(1/(GAMMADIST(1,J6,1,0)*EXP(1))))/0.000001–LN(G6/H6)

J6: Use SOLVER or GOAL SEEK to find alpha to make the equation in I6 zero

K6=J6/G6 (alpha divided by the average)

L6=[100*J6*LN(K6)–100*LN(1/(GAMMADIST(1,J6,1,0)*EXP(1))+(J6-1)*SUM(LN(B$6:B$105)) +100*K6*D6–K6*SUM(B$6:B$105)] (array summation)

TABLE 2.8

Control Limits and Center Line for
Three-Parameter Gamma Distribution
Critical Values for Col_1

Lower Tail Area (<=)	Gamma (3-Parameter)
0.00135	1.26538
0.1	2.44861
0.5	4.00592
0.9	6.27442
0.99865	10.5841

Spreadsheet Deployment, Chart for Nonnormal Individuals

Table 2.9 shows part of a Microsoft Excel spreadsheet that displays a control chart (Figure 2.11) for the points from the simulation under discussion (100 random points from a gamma distribution with $\alpha = 8$, $\gamma = 2$).

The conditional formatting of cells B8 and below in Table 2.9 is as shown in Figure 2.10. If the operator enters a value for X that is less than the LCL (cell C8) or more than the UCL (cell E8), the cell background will turn red. ISNUMBER(B8) prevents interpretation of a blank cell as a zero. The operator

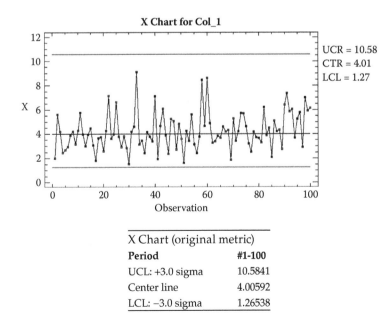

X Chart for Col_1

UCR = 10.58
CTR = 4.01
LCL = 1.27

X Chart (original metric)

Period	#1-100
UCL: +3.0 sigma	10.5841
Center line	4.00592
LCL: −3.0 sigma	1.26538

FIGURE 2.9A
StatGraphics control chart for three-parameter gamma distribution.

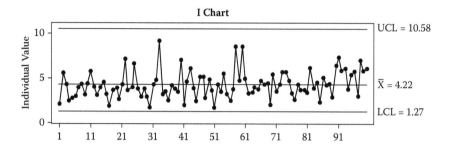

FIGURE 2.9B
Minitab control chart for 3-parameter gamma distribution.

will not even have to look at the control chart to know that the measurement is outside the control limits.

It is in fact possible to determine the status of each measurement without separate columns for control limits, and without an actual control chart, by means of the conditional format in Figure 2.12. The conditional format statement calculates the quantile of the measurement (or statistic, such as sample average and range) and changes the cell color if this quantile is outside an acceptable range.

It was not possible to copy a screenshot of the complete conditional test for cell B12, which is:

$$IF(OR(GAMMADIST(B12-\$C\$6,\$C\$4,1/\$C\$5,1)<0.00135,$$

$$GAMMADIST(B12-\$C\$6,\$C\$4,1/\$C\$5,1)>0.99865)*ISNUMBER(B12),1)$$

TABLE 2.9

Spreadsheet, Chart for Nonnormal Individuals

	A	B	C	D	E
3	UCL	10.5841			
4	CL	4.00592			
5	LCL	1.26538			
6					
7	**Sample**	**X**	**LCL**	**CL**	**UCL**
8	1	2.005	1.265	4.006	10.584
9	2	5.585	1.265	4.006	10.584
10	3	4.177	1.265	4.006	10.584
	4	2.465	1.265	4.006	10.584
	5	2.692	1.265	4.006	10.584
	6	2.926	1.265	4.006	10.584
	7	3.860	1.265	4.006	10.584
	8	4.210	1.265	4.006	10.584
	9	3.145	1.265	4.006	10.584
	10	4.276	1.265	4.006	10.584

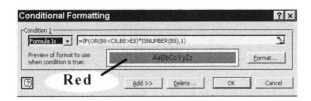

FIGURE 2.10
Conditional formatting of cells to show out-of-control conditions.

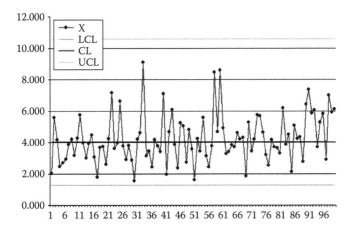

FIGURE 2.11
Excel control chart for nonnormal individuals.

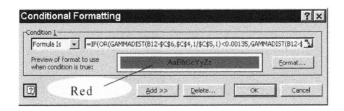

FIGURE 2.12
Computed visual control for in- or out-of-control status.

where $C$6 is the threshold parameter to be deducted from the measurement in cell B12 before calculation of the cumulative gamma distribution. $C$4 is the shape parameter and $C$5 is the scale parameter; remember that Excel's GAMMADIST uses $1/\gamma$ as the function's argument. The requirement that ISNUMBER(B12) be true prevents interpretation of a blank cell as zero. If the CDF of the entry is less than 0.00135 or greater than 0.99865 (in practice, only the upper limit is generally of practical interest for the gamma distribution) the cell background turns red. This kind of *visual control* will work for almost any control chart, including the range charts in Chapter 3.

This chapter has shown so far how to fit the two-parameter and three-parameter gamma distributions to process data, and how to set up control charts for individual measurements. The next section will address the treatment of subgroups.

Subgroup Averages of the Gamma Distribution

This section will show that the average of n measurements from a gamma distribution with parameters (α,γ,δ) follows another gamma distribution with parameters $(n\alpha,n\gamma,\delta)$. Recall the procedure for transformation of the joint probability distribution for two random variables:

$$g(u,v) = f(x(u,v), y(u,v))J\left(\frac{x,y}{u,v}\right) \quad \text{where} \quad J\left(\frac{x,y}{u,v}\right) = \begin{vmatrix} \dfrac{\partial x}{\partial u} & \dfrac{\partial x}{\partial v} \\ \dfrac{\partial y}{\partial u} & \dfrac{\partial y}{\partial v} \end{vmatrix}$$

In this case, develop the joint probability of getting two measurements u and v from a two-parameter gamma distribution.

$$h(x) = \frac{\gamma^\alpha}{\Gamma(\alpha)} x^{\alpha-1} \exp(-\gamma x) \quad \text{and} \quad h(y) = \frac{\gamma^\alpha}{\Gamma(\alpha)} y^{\alpha-1} \exp(-\gamma y)$$

$$f(x,y) = \frac{\gamma^{2\alpha}}{\Gamma^2(\alpha)} (xy)^{\alpha-1} \exp(-\gamma(x+y))$$

$$u(x,y) = \frac{x+y}{2} \quad \text{and} \quad v(x,y) = y$$

$$\Rightarrow x = 2u - v \quad \text{and} \quad y = v$$

$$\Rightarrow J\left(\frac{x,y}{u,v}\right) = \begin{vmatrix} 2 & -1 \\ 0 & 1 \end{vmatrix} = 2$$

$$\Rightarrow g(u,v) = \frac{\gamma^{2\alpha}}{\Gamma^2(\alpha)} (2u-v)^{\alpha-1} v^{\alpha-1} \exp(-\gamma(2u-v+v)) \times 2$$

The next step is to get the *marginal distribution* for u by performing the integration $\int_0^\infty g(u,v)dv$. This requires rearrangement of $g(u,v)$ into the *beta distribution* for which this integral is defined.

$$g(u,v) = 2\frac{\gamma^{2\alpha}}{\Gamma^2(\alpha)}(2u-v)^{\alpha-1}v^{\alpha-1}\exp(-2u\gamma)$$

$$= 2\frac{\gamma^{2\alpha}}{\Gamma^2(\alpha)}\exp(-2u\gamma)(2u-v)^{\alpha-1}v^{\alpha-1}$$

$$= 2\frac{\gamma^{2\alpha}}{\Gamma^2(\alpha)}\exp(-2u\gamma)(2u)^{2\alpha-2}\left(1-\frac{v}{2u}\right)^{\alpha-1}\left(\frac{v}{2u}\right)^{\alpha-1}$$

$$\beta(q,s) = \int_0^\infty w^{q-1}(1+w)^{-(q+s)}\,dw = \int_0^1 (1-p)^{q-1}p^{s-1}\,dp$$

where $p = \dfrac{1}{1+w}$ and $w = \dfrac{1}{p}-1$

Let $f(u)$ now refer to the marginal distribution for the average, and let q and s equal the shape parameter α. Note in the integration limit that v cannot be greater than $2u$ because of $(1-\frac{v}{2u})^{\alpha-1}$.

$$f(u) = \int_0^{2u} = 2\frac{\gamma^{2\alpha}}{\Gamma^2(\alpha)}\exp(-2u\gamma)(2u)^{2\alpha-2}\left(1-\frac{v}{2u}\right)^{\alpha-1}\left(\frac{v}{2u}\right)^{\alpha-1}dv$$

Let $p = \dfrac{v}{2u} \Rightarrow dv = 2u\,dp$

Also note that $\left(1-\dfrac{v}{2u}\right)^{\alpha-1} \Rightarrow 0 \le v \le 2u \Rightarrow p \in [0,1]$

$$f(u) = \int_0^1 2\frac{\gamma^{2\alpha}}{\Gamma^2(\alpha)}\exp(-2u\gamma)(2u)^{2\alpha-2}p^{\alpha-1}(1-p)^{\alpha-1}2u\,dp$$

$$= 4u\frac{\gamma^{2\alpha}}{\Gamma^2(\alpha)}\exp(-2u\gamma)(2u)^{2\alpha-2}\int_0^1 p^{\alpha-1}(1-p)^{\alpha-1}\,dp$$

$$= 2\frac{\gamma^{2\alpha}}{\Gamma^2(\alpha)}\exp(-2u\gamma)(2u)^{2\alpha-1}\frac{\Gamma(\alpha)\Gamma(\alpha)}{\Gamma(2\alpha)}$$

$$= \frac{(2\gamma)^{2\alpha}}{\Gamma(2\alpha)}u^{2\alpha-1}\exp(-2\gamma u)$$

The result is the expression for the average of two measurements from the gamma distribution with shape parameter 2α and scale parameter 2γ, as shown in Equation Set 2.7.

$$f\left(u = \frac{x+y}{2}\right) = \frac{\gamma^{2\alpha}}{\Gamma(2\alpha)} \exp(-2\gamma u)(2u)^{2\alpha-2}$$

(Set 2.7)

$$\mu = \frac{2\alpha}{2\gamma} = \frac{\alpha}{\gamma} \quad \text{and} \quad \sigma^2 = \frac{2\alpha}{(2\gamma)^2} = \frac{\alpha}{2\gamma^2}$$

The average is the average of the gamma distribution, and the variance is as expected for the average of two numbers. By extension of Equation Set 2.7, we derive Equation Set 2.8.

$$f\left(u = \frac{1}{n}\sum_{i=1}^{n} x_i\right) = \frac{\gamma^{n\alpha}}{\Gamma(n\alpha)} \exp(-n\gamma u)(nu)^{n(\alpha-1)}$$

(Set 2.8)

$$\mu = \frac{n\alpha}{n\gamma} = \frac{\alpha}{\gamma} \quad \text{and} \quad \sigma^2 = \frac{n\alpha}{(n\gamma)^2} = \frac{\alpha}{n\gamma^2}$$

Consider 100 samples of 4 from a gamma distribution with shape parameter $\alpha = 8$, scale parameter $\gamma = 2$, and threshold parameter $\delta = 4$. The averages should have a shape parameter of 32, a scale parameter of 8, and a threshold parameter of 4. The average of the 100 averages is 8.03 (8/2 + 4) and their variance is 0.441, which is close to the expected value of 0.5.

Minitab, StatGraphics, and the MathCAD bisection routine were, however, unable to fit the simulated data properly. The former yielded negative threshold parameters, and the latter's graph of the likelihood function versus δ had a maximum at zero with a negative slope, an implication that the optimum threshold parameter is negative. Part of the problem may be the fact that the resulting distribution is almost indistinguishable from a normal distribution because of the central limit theorem.

A gamma distribution should in fact become more normal as the shape parameter increases because it represents the sum of α measurements from an exponential distribution with scale parameter γ (Mood, Graybill, and Boes, 1974, 193). The average of 32 such measurements should therefore behave as if it comes from a normal distribution.

In any event, it is better to let the computer fit the distribution from the individual values instead of the averages. The resulting parameters may then be used to set the control limits for the x-bar chart. Always remember that the central limit theorem works for control charts if the sample is sufficiently large, but it does not work for estimates of the nonconforming fraction.

Average Run Length (ARL), Gamma Distribution

The ARL for the x-bar chart for the normal distribution is very straight-forward because it is a direct function of changes in a single distribution parameter. The mean of the gamma distribution is, however, a function of two parameters (and three if the threshold parameter is not held constant). Calculation of the ARL should therefore assume that only one parameter changes while the other remains constant.

The gamma distribution is the continuous scale analogue of the Poisson distribution, and its scale parameter corresponds to the Poisson distribution's random arrival rate. Assume that an undesirable process change, such as an increase in impurity levels, defects, or particle counts, corresponds to a change in the scale parameter while the shape parameter remains constant. Then the Type II risk that an individual measurement will be below the upper control limit is $\beta(\gamma) = \Pr(X \le UCL)$ where UCL was computed on the basis of the parameters that were fitted to the data. Then

$$ARL = \frac{1}{1 - F(UCL|\gamma)} = \frac{1}{1 - \int_0^{UCL} \frac{\gamma^\alpha}{\Gamma(\alpha)}(x - \delta)^{\alpha-1}\exp(-\gamma(x - \delta))} \qquad (2.9)$$

where α and δ remain constant. Similar calculations can be performed for the sample average and for changes in the shape parameter.

The Weibull Distribution

The Weibull distribution appears frequently in reliability engineering, where it is often a good model for time to failure or cycles to failure. It is a trans-formation of the extreme value distribution, which models applications in which failure takes place at the weakest point. It was found (Levinson, 1999a) to be a good model for the force necessary to cause shear failure in epoxy cement that bonds semiconductor chips to ceramic circuit boards.

The three-parameter Weibull probability density function and cumulative density functions are as follows, where β is the shape parameter, θ is the scale parameter, and δ is the threshold parameter. (Some references use λ, the reciprocal of θ, as the scale parameter, and it corresponds to the hazard rate of an exponential distribution.)

$$f(x) = \frac{\beta}{\theta}\left(\frac{x - \delta}{\theta}\right)^{\beta-1}\exp\left(-\left(\frac{x - \delta}{\theta}\right)^\beta\right)$$

$$F(x) = 1 - \exp\left(-\left(\frac{x - \delta}{\theta}\right)^\beta\right)$$

(Set 2.10)

The log likelihood function is (Lawless, 1982, 170):

$$\ln L(X) = \sum_{i=1}^{n} \ln\beta - \beta\ln\theta + (\beta-1)\ln(x_i - \delta) - \left(\frac{x_i - \delta}{\theta}\right)^{\beta}$$

$$= n\ln\beta - n\beta\ln\theta + \sum_{i=1}^{n}(\beta-1)\ln(x_i - \delta) - \sum_{i=1}^{n}\left(\frac{x_i - \delta}{\theta}\right)^{\beta}$$

(Set 2.11)

Solution of the partial derivatives for β and θ yields equations whose roots yield the optimal parameters. Begin with the scale parameter:

$$\frac{\partial \ln L(x)}{\partial \theta} = -\frac{n\beta}{\theta} + \frac{\beta}{\theta^{\beta+1}}\sum_{i=1}^{n}(x_i - \delta)^{\beta} = 0$$

$$\Rightarrow \theta^{\beta} = \frac{1}{n}\sum_{i=1}^{n}(x_i - \delta)^{\beta} \quad \text{and} \quad \theta = \left(\frac{1}{n}\sum_{i=1}^{n}(x_i - \delta)^{\beta}\right)^{\frac{1}{\beta}}$$

Solution for the shape parameter:

$$\frac{\partial \ln L(x)}{\partial \beta} = \frac{n}{\beta} - n\beta\ln\theta + \sum_{i=1}^{n}\ln(x_i - \delta) - \sum_{i=1}^{n}\left(\frac{x_i - \delta}{\theta}\right)^{\beta}\ln\left(\frac{x_i - \delta}{\theta}\right)$$

$$\theta = \left(\frac{1}{n}\sum_{i=1}^{n}(x_i - \delta)^{\beta}\right)^{\frac{1}{\beta}} \Rightarrow \ln\theta = \frac{1}{\beta}\ln\sum_{i=1}^{n}(x_i - \delta)^{\beta} - \frac{\ln n}{\beta}$$

and then

$$\ln L(X) = n\ln\beta - n\beta\left(\frac{1}{\beta}\ln\sum_{i=1}^{n}(x_i - \delta)^{\beta} - \frac{\ln n}{\beta}\right)$$

$$\cdots + \sum_{i=1}^{n}(\beta-1)\ln(x_i - \delta) - \frac{n\sum_{i=1}^{n}(x_i - \delta)^{\beta}}{\sum_{i=1}^{n}(x_i - \delta)^{\beta}}$$

noting that $\theta^{-\beta} = \left(\frac{1}{n}\sum_{i=1}^{n}(x_i - \delta)^{\beta}\right)^{-1}$

$$\Rightarrow \ln L(X) = n\ln\beta - n\ln\sum_{i=1}^{n}(x_i - \delta)^{\beta} - n\ln n + \sum_{i=1}^{n}(\beta-1)\ln(x_i - \delta) - n$$

Now take the partial derivative for β:

$$\frac{\partial \ln L(X)}{d\beta} = \frac{n}{\beta} - n\frac{\sum_{i=1}^{n} \ln(x_i - \delta)(x_i - \delta)^{\beta}}{\sum_{i=1}^{n}(x_i - \delta)^{\beta}} + \sum_{i=1}^{n} \ln(x_i - \delta) = 0$$

$$\Rightarrow \frac{\sum_{i=1}^{n} \ln(x_i - \delta)(x_i - \delta)^{\beta}}{\sum_{i=1}^{n}(x_i - \delta)^{\beta}} - \frac{1}{\beta} - \sum_{i=1}^{n} \ln(x_i - \delta) = 0$$

For any given threshold parameter and n measurements,

$$\text{Solve } \frac{\sum_{i=1}^{n}(x_i - \delta)^{\beta} \ln(x_i - \delta)}{\sum_{i=1}^{n}(x_i - \delta)^{\beta}} - \frac{1}{\beta} - \frac{1}{n}\sum_{i=1}^{n} \ln(x_i - \delta) = 0$$

$$\text{(Set 2.12)}$$

$$\text{and then } \theta = \left(\frac{1}{n}\sum_{i=1}^{n}(x_i - \delta)^{\beta} \right)^{\frac{1}{\beta}}$$

As with the gamma distribution, it is necessary to select the threshold parameter (if any) that maximizes the likelihood function or its natural log. The procedure is similar to that for the gamma distribution, where the natural log of the maximum likelihood function is maximized as a function of the threshold parameter.

Example: Two-Parameter Weibull Distribution

Two hundred random data were simulated from a Weibull distribution with shape parameter $\beta = 2$ and scale parameter $\theta = 0.5$ (Weibull_b2_theta_0.5 in the simulated data spreadsheet). Figure 2.13 illustrates a bisection routine that optimizes the parameters for the data, and the results appear at the bottom.

The Visual Basic for Applications function FitWeibull.bas uses a bisection routine similar to that for FitGamma.bas to obtain $\beta = 1.9686$ and $\theta = 0.48214$. Its arguments are the range of data in the spreadsheet and the maximum allowable shape parameter. If the user does not specify a sufficiently high shape parameter, the function will indicate that it cannot find a solution. Minitab's Weibull process capability analysis yields the same results, as does StatGraphics if the results are rounded to the nearest ten-thousandth.[5]

Figure 2.14 shows the histogram and quantile-quantile plot from StatGraphics. Again, these are things the practitioner should expect to see in capability reports from suppliers, along with the chi square test for goodness of fit.

The VBA function QQ_Plot generates a quantile-quantile plot with a slope of 1.021, intercept −0.0102, and correlation 0.996. The format of the command is   =QQ_Plot(F6:I55,"Weibull",1.9868,0.48214,0,"C:\TEMP\QQ_Weibull.txt"). =ChiSquare_Weibull(F6:I55,16,1.9686,0.48214,0,"C:\TEMP\ChiSquare_Weibull. txt") begins with 16 cells and, after combination of the tail cells to meet the requirement that the expected count be 5 or greater, yields a chi square test statistic of 10.94 for 12 cells and therefore 9 degrees of freedom. The resulting

$$n: = \text{rows}(x) \qquad x \text{ is the data vector}$$

$$f(\beta):= \frac{\sum\limits_{i=1}^{n}(x_i)^{\beta}.\ln(x_i)}{\sum\limits_{i=1}^{n}(x_i)^{\beta}} - \frac{1}{\beta} - \frac{1}{n}.\sum\limits_{i=1}^{n}\ln(x_i) \qquad \text{Function to be driven to zero}$$

Range to bisect $\begin{bmatrix} a \\ b \end{bmatrix} := \begin{bmatrix} 1 \\ 10 \end{bmatrix}$ $\qquad f(a).f(b) = -0.662 \qquad$ Test to make sure a solution is in the range to be bisected

Graph of the function $\qquad\qquad j:= 1..101 \qquad y_j := a + \frac{(j-1)}{100}.(b-a)$

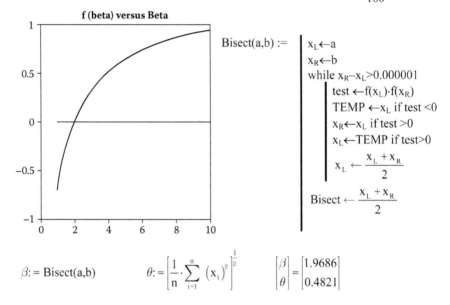

$$\beta := \text{Bisect}(a,b) \qquad \theta := \left[\frac{1}{n}\cdot\sum_{i=1}^{n}(x_i)^{\beta}\right]^{\frac{1}{\beta}} \qquad \begin{bmatrix}\beta \\ \theta\end{bmatrix} = \begin{bmatrix}1.9686 \\ 0.4821\end{bmatrix}$$

FIGURE 2.13
MathCAD bisection routine, two-parameter Weibull distribution.

table appears below the simulated dataset in tab Weibull_b2_theta_0.5 of the simulated data spreadsheet.

Calculation of the quantiles for a control chart is extremely straightforward for this distribution:

$$1-F(x)=\exp\left(-\left(\frac{x-\delta}{\theta}\right)^{\beta}\right) \Rightarrow x = \theta(-\ln(1-F(x)))^{1/\beta}+\delta$$

(Set 2.13)

or $F^{-1}(q) = \theta(-\ln(1-q))^{1/\beta}+\delta$

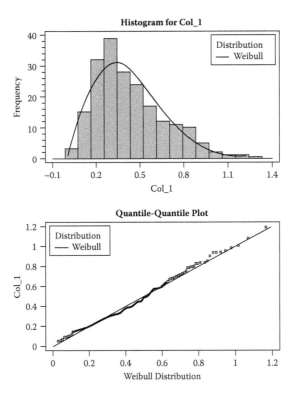

FIGURE 2.14
Histogram and quantile-quantile plot, Weibull distribution.

The center line of the chart for individuals is $0.48214(-\ln(1-0.5))^{\frac{1}{1.9686}} = 0.4002$, and the procedure is similar for the Shewhart-equivalent control limits at the 0.00135 and 0.99865 quantiles. Figure 2.15 shows the resulting control chart from StatGraphics.

Calculation of the nonconforming fractions, and then of the normal-equivalent process performance indices, is equally straightforward whether or not a threshold parameter is present:

$$F(LSL) = 1 - \exp\left(-\left(\frac{LSL-\delta}{\theta}\right)^{\beta}\right) \text{ and } 1 - F(USL) = \exp\left(-\left(\frac{USL-\delta}{\theta}\right)^{\beta}\right)$$

In this case, suppose the specification limits are [0.02,1.30]. Then

$$F(0.02) = 1 - \exp\left(-\left(\frac{0.02}{0.48214}\right)^{1.9686}\right) = 0.001900$$

$$1 - F(1.30) = \exp\left(-\left(\frac{1.30}{0.48214}\right)^{1.9686}\right) = 0.0008700$$

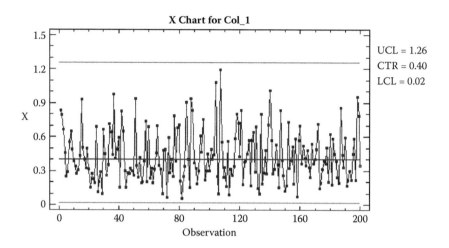

FIGURE 2.15
X chart for a Weibull distribution.

The process performance indices are then

$$PPL = \frac{-\Phi^{-1}(0.00190)}{3} = 0.965$$

and

$$PPU = \frac{-\Phi^{-1}(0.000870)}{3} = 1.044.$$

StatGraphics obtains the results in Table 2.10, and the process performance indices as calculated above are the long-term performance indices.

TABLE 2.10

StatGraphics Process Capability Analysis, Weibull Distribution
    Specifications
    USL = 1.3
    LSL = 0.02

	Short-Term Capability	Long-Term Performance
Sigma (after normalization)	1.01041	1.0
C_p/P_p	0.993901	1.00425
C_{pk}/P_{pk}	0.954804	0.964741
C_{pk}/P_{pk} (upper)	1.033	1.04375
C_{pk}/P_{pk} (lower)	0.954804	0.964741
DPM	3059.94	2770.85

Note: Based on 6.0 sigma limits in the normalized metric. Short-term sigma estimated from average moving range.

Figure 2.16 shows the results from Minitab.

The next step is to deal with the three-parameter Weibull distribution. The procedure is very similar to that for the three-parameter gamma distribution.

Three-Parameter Weibull Distribution

As with the three-parameter gamma distribution, the shape and scale parameters will depend on the selected threshold parameter. The idea is to find the threshold parameter that maximizes the likelihood function or its log. Given 200 simulated data from a distribution with $\beta = 1.5$, $\theta = 2$, and $\delta = 1$ (Weibull_b1.5_theta_2_delta_1 in the simulated data file), Figure 2.17 shows the solution procedure in MathCAD. The graph of the log likelihood function versus the threshold parameter illustrates the optimization of the threshold parameter, which is then handled analytically by a bisection routine. The best fit from the MathCAD program is $\beta = 1.5852$, $\theta = 1.7876$, $\delta = 1.0083$, and StatGraphics Centurion delivers identical results.

The user disk includes the VBA function FitWeibull3.bas, which returns $\beta = 1.5852$, $\theta = 1.7875$, and $\delta = 1.0083$. Function ChiSquareWeibull.bas returns a chi square test statistic of 8.10 for eleven cells.

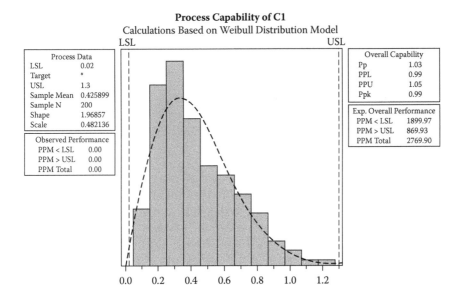

FIGURE 2.16
Minitab process capability analysis, Weibull distribution.

Three-Parameter Weibull Distribution   ORIGIN:=1
Initial statistics

$$\begin{bmatrix} n \\ \delta_{min} \\ \delta_{max} \end{bmatrix} := \begin{bmatrix} rows(x) \\ 0 \\ min(x) \cdot 0.9999 \end{bmatrix} \qquad \begin{bmatrix} n \\ \delta_{min} \\ \delta_{max} \end{bmatrix} = \begin{bmatrix} 200 \\ 0 \\ 1.028 \end{bmatrix}$$

The maximum allowable threshold parameter must be slightly less than the minimum observation in the data set because a calculation involves the natural log of x minus the threshold parameter.

Preliminary graph of the log likelihood function versus threshold parameter

$$\ln\_Weibull\_pdf(y, \beta, \theta, \delta) := \ln(\beta) - \beta \cdot \ln(\theta) + (\beta-1) \cdot \ln(y - \delta) - \left(\frac{y - \delta}{\theta}\right)^{\beta}$$

Equation to be solved
for the shape parameter
$$f(\beta, \delta) := \frac{\sum_{i=1}^{n}(x_i - \delta)^{\beta} \cdot \ln(x_i - \delta)}{\sum_{i=1}^{n}(x_i - \delta)^{\beta}} - \frac{1}{\beta} - \frac{1}{n} \cdot \sum_{i=1}^{n} \ln(x_i - \delta)$$

FIGURE 2.17A
Fit of three-parameter Weibull distribution (MathCAD), part 1.

Calculation of the Shewhart-equivalent control limits for the X chart is straightforward:

$$F^{-1}(q) = \theta(-\ln(1-q))^{1/\beta} + \delta \Rightarrow$$

$$F^{-1}(0.00135) = 1.7875(-\ln(1-0.00135))^{1/1.5852} + 1.0083 = 1.036$$

$$F^{-1}(0.99865) = 1.7875(-\ln(1-0.99865))^{1/1.5852} + 1.0083 = 6.891$$

and these are in fact the limits found by Minitab 15, as shown in Figure 2.18.

Samples from the Weibull Distribution

The average of measurements from a Weibull distribution is an unresolved problem because in contrast to the gamma distribution, there seems to be no way to derive an equation for it.[6] Lawless (1982, 195) adds that "life test plans under the Weibull model have not been very thoroughly investigated, however, because of the complexity of the associated distributional problems. ... it is almost always impossible to determine exact small-sample properties or to make effective comparisons of plans, except by simulation." We propose the following as what might be the best available solution.

$\ln L(d) :=$ | $x_L \leftarrow 1$ The maximum allowable shape

$x_R \leftarrow 5$ parameter is 5; the user can change

while $x_R - x_L > 0.000001$ this if necessary.

$$\text{test} \leftarrow f(x_L, d) \cdot f(x_R, d)$$

$$\text{TEMP} \leftarrow x_L \text{ if } \text{test} < 0$$

$$x_R \leftarrow x_L \text{ if } \text{test} > 0$$

$$x_L \leftarrow \text{TEMP if } \text{test} > 0$$

$$x_L \leftarrow \frac{x_L + x_R}{2}$$

$$\beta \leftarrow \frac{x_L + x_R}{2}$$

This part of the algorithm finds the shape and scale parameters that go with the threshold parameter in the function argument.

$$\theta \leftarrow \left[\frac{1}{n} \cdot \sum_{i=1}^{n} (x_i - d)^\beta \right]^{\frac{1}{\beta}}$$

$$\ln L \leftarrow \sum_{i=1}^{n} \ln\_\text{Weibull\_pdf}(x_i, \beta, \theta, d)$$

FIGURE 2.17B
Fit of three-parameter Weibull distribution, part 2.

A chart for process variation is available for the Weibull distribution, and Chapter 3 will show how to construct it. The corresponding X (individuals) chart can display the individual measurements separately and flag any that exceed the control limits for individuals. Figure 2.19 shows how the two charts (using Weibull_b1.5_theta_2_delta_1 from the simulated datasets) might be displayed in tandem.

For any given sample of n pieces, the false alarm risk of the individuals chart is $1 - (1 - \alpha)^n$ where α is the risk for a single individual, that is, the same risk for n points on a traditional X chart. The Type II risk for any given change in the process parameters is $F^n(UCL \mid \text{new\_parameters})$, that is, the chance that all n measurements will be below the upper control limit *given* the new set of parameters. For the lower control limit, it is $(1 - F(LCL \mid \text{new\_parameters}))^n$.

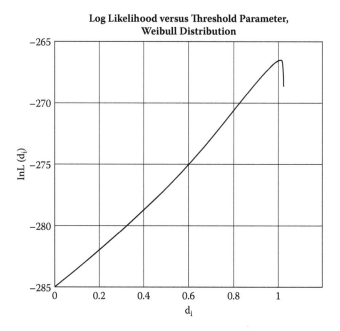

FIGURE 2.17C
Fit of three-parameter Weibull distribution, part 3.

The power of this chart to detect undesirable process changes is almost certainly less than that of a chart for sample averages if one could be developed, and this treatment of samples from the Weibull distribution is a second-best alternative.

The Lognormal Distribution

Data follow a lognormal distribution if the natural logs of the measurements follow the normal distribution. This allows direct use of ln(x) on control charts and for capability indices; as an example,

$$PPU = \frac{(\ln(USL) - \mu)}{3\sigma_p}$$

where μ is the mean natural log and σ_p is the standard deviation that is calculated from the natural logs.

Lawless (1982, 24) reports that this distribution has been applied successfully to failure times for electrical insulation and onset times for lung cancer

Bisect the interval for the point at which the slope is zero (maximum)

$$\Delta := 0.000001 \qquad d\_lnL(d) := \frac{lnL(d+\Delta) - lnL(d)}{\Delta}$$

Numerical estimate for the

first derivative at any point

$$d\_ML(a, b) :=$$

$\quad | \quad x_L \leftarrow a$

$\quad | \quad x_R \leftarrow b$

$\quad | \quad$ while $x_R - x_L > 0.000001$ The objective is to find the

$\quad | \quad\quad | \quad$ test $\leftarrow d\_lnL(x_L) \cdot d\_lnL(x_R)$ threshold parameter at

$\quad | \quad\quad | \quad$ TEMP $\leftarrow x_L$ if test < 0 which the first derivative

$\quad | \quad\quad | \quad$ $x_R \leftarrow x_L$ if test > 0 is zero.

$\quad | \quad\quad | \quad$ $x_L \leftarrow$ TEMP if test > 0 $\delta := d\_ML(\delta_{min}, \delta_{max})$

$\quad | \quad\quad | \quad$ $x_L \leftarrow \dfrac{x_L + x_R}{2}$ $\delta = 1.0083$

$\quad | \quad$ d$\_$ML $\leftarrow \dfrac{x_L + x_R}{2}$

Now report the shape and scale parameters for this threshold parameter

$$\text{Bisect}(a, b) :=$$

$\quad | \quad x_L \leftarrow a$

$\quad | \quad x_R \leftarrow b$

$\quad | \quad$ while $x_R - x_L > 0.000001$ $\beta := \text{Bisect}(1,5)$

$\quad | \quad\quad | \quad$ test $\leftarrow f\_(x_L, \delta) \cdot f(x_R, \delta)$

$\quad | \quad\quad | \quad$ TEMP $\leftarrow x_L$ if test < 0 $\theta := \left[\dfrac{1}{n} \cdot \sum\limits_{i=1}^{n} (x_i - \delta)^\beta \right]^{\frac{1}{\beta}}$

$\quad | \quad\quad | \quad$ $x_R \leftarrow x_L$ if test > 0

$\quad | \quad\quad | \quad$ $x_L \leftarrow$ TEMP if test > 0

$\quad | \quad\quad | \quad$ $x_L \leftarrow \dfrac{x_L + x_R}{2}$ $\begin{bmatrix} \beta \\ \theta \\ \delta \end{bmatrix} = \begin{bmatrix} 1.58522 \\ 1.78757 \\ 1.00831 \end{bmatrix}$

$\quad | \quad$ Bisect $\leftarrow \dfrac{x_L + x_R}{2}$

FIGURE 2.17D
Fit of three-parameter Weibull distribution, part 4.

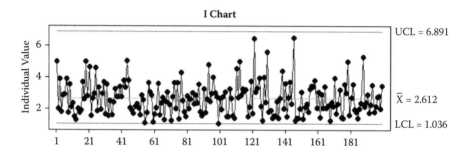

FIGURE 2.18
Minitab capability six pack control chart, three-parameter Weibull.

in cigarette smokers. The reference adds that it is a questionable model for reliability applications because the hazard function increases, reaches a maximum, and then decreases. This is the opposite of the "bathtub model" in which the hazard rate decreases as defective items fail, runs flat, and then increases as the surviving population wears out.

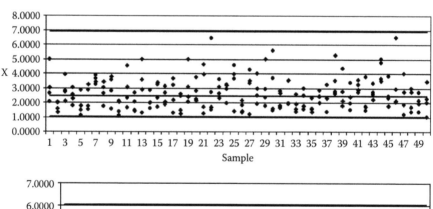

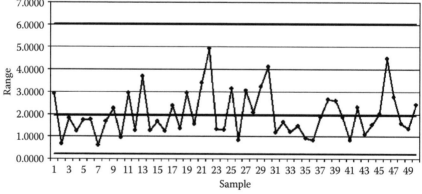

FIGURE 2.19
Individuals and range chart, Weibull distribution.

Measurements with Detection Limits (Censored Data)

The issue of data censoring applies primarily to reliability applications in which the time, mileage, cycles, or other units of use at which the item fails are the response variable of primary interest. There are cases in which not all units can be tested to failure, and this is the setting for *right-censoring*. In Type I censoring, which is of primary interest here, the test must be discontinued after a certain amount of time or other units of use. In Type II censoring, the test is stopped after r out of n units fail.

Note that in both cases, only r out of n units fail the reliability test; the difference consists of the stopping criterion. In Type I censoring, the stopping criterion is time (e.g., "Run the test for 200 hours and assess the results."). In Type II censoring, it is the number of failures (e.g., "Run the test until 90 out of the 100 specimens fail."). Lawless (1982) shows how to handle right-censored data for the gamma and Weibull distributions.

Right-censoring is not particularly applicable to statistical process control because the measurements rarely involve testing until failure. *Left-censoring* sometimes occurs for measurements of trace impurities when the instrumentation cannot detect less than a given quantity. Suppose, for example, that the instrument cannot detect fewer than 10 parts per billion of the impurity in question. Any measurement of 10 ppb (or possibly a measurement that is reported as zero) really means "10 ppb or less," and the Environmental Protection Agency refers to such measurements as "non-detects." Maximum likelihood estimation methods can be used for either type of censoring as follows:

$$\text{Maximize} \quad \ln L(X|\theta) = \sum_{i=1}^{n} \ln l(x_i \mid \theta)$$

where θ is the set of distribution parameters

$$l(x_i|\theta) = \begin{cases} f(x_i|\theta) & \text{if uncensored} \\ F(x_i|\theta) & \text{if left-censored} \\ 1 - F(x_i|\theta) & \text{if right-censored} \end{cases}$$

Note that in the cases of right-censoring, x_i can be the fixed time at which the test ends. In the case of left-censoring, it can be the lower detection limit. The chapter will address the simpler Weibull distribution first, although the gamma distribution is more likely to apply to these situations.

Left-Censored Weibull Distribution

A lower detection limit generally suggests that the quality characteristic can be zero, and there is no need to worry about threshold parameters. The two-parameter Weibull distribution will therefore apply:

$$f(x) = \frac{\beta}{\theta}\left(\frac{x}{\theta}\right)^{\beta-1} \exp\left(-\left(\frac{x}{\theta}\right)^{\beta}\right) \quad \text{and} \quad F(x) = 1 - \exp\left(-\left(\frac{x}{\theta}\right)^{\beta}\right)$$

In the right-censored case, the survivor function simply becomes part of the basic equation, and subsequent treatment of the log likelihood function differs little from the situation in which there is no censoring. Given r failures out of n pieces,

$$L(X) = \prod_{i=1}^{r} \frac{\beta}{\theta}\left(\frac{x_i}{\theta}\right)^{\beta-1} \exp\left(-\left(\frac{x}{\theta}\right)^{\beta}\right) \times \prod_{j=1}^{n-r} \exp\left(-\left(\frac{x_j = T_{end}}{\theta}\right)^{\beta}\right)$$

$$nL(X) = \sum_{i=1}^{r} \ln\beta - \beta\ln\theta + (\beta-1)\ln(x_i) - \left(\frac{x_i}{\theta}\right)^{\beta} + (n-r)\left(-\left(\frac{x_j = T_{end}}{\theta}\right)^{\beta}\right)$$

$$= r\ln\beta - r\beta\ln\theta + (\beta-1)\sum_{i=1}^{r} \ln(x_i) - \sum_{i=1}^{n}\left(\frac{x_i}{\theta}\right)^{\beta}$$

where the last term in the equation uses

$$x_i = x_i \quad \text{if failed}$$

$$x_i = T_{end} \quad \text{if survived}$$

The left-censored case is more complicated because it uses the cumulative density function as opposed to the survivor function, and the equation

$$\frac{d\ln L(x)}{d\theta} = 0$$

has no closed form as it does in the uncensored or right-censored cases. Suppose that r out of n measurements exceed the detection limit D (they are not censored) and $n - r$ measurements are reported at the detection limit D (are censored):

$$L(X) = \prod_{i=1}^{r} \frac{\beta}{\theta}\left(\frac{x_i}{\theta}\right)^{\beta-1} \exp\left(-\left(\frac{x_i}{\theta}\right)^{\beta}\right) \times \prod_{j=1}^{n-r}\left(1 - \exp\left(-\left(\frac{D}{\theta}\right)^{\beta}\right)\right)$$

The log likelihood function is then:

$$\ln L(X) = \sum_{i=1}^{r} \ln \beta - \beta \ln \theta + (\beta - 1)\ln(x_i) - \left(\frac{x_i}{\theta}\right)^{\beta}$$

$$+ (n-r)\ln\left(1 - \exp\left(-\left(\frac{D}{\theta}\right)^{\beta}\right)\right)$$

$$= r \ln \beta - r\beta \ln \theta + \sum_{i=1}^{r} (\beta - 1)\ln(x_i) - \sum_{i=1}^{r}\left(\frac{x_i}{\theta}\right)^{\beta}$$

$$+ (n-r)\ln\left(1 - \exp\left(-\left(\frac{D}{\theta}\right)^{\beta}\right)\right)$$

Then

$$\frac{\partial \ln L(X)}{\partial \theta} = -\frac{r\beta}{\theta} + \frac{\beta}{\theta^{\beta+1}}\sum_{i=1}^{r}(x_i - \delta)^{\beta}$$

$$+ (n-r)\beta\frac{D}{\theta^2\left(\exp\left(-\left(\frac{D}{\theta}\right)^{\beta}\right)-1\right)}\exp\left(-\left(\frac{D}{\theta}\right)^{\beta}\right)$$

which has no closed form solution for θ when set equal to zero. The situation for the gamma distribution is just as bad, and some kind of numerical solution may be required (noting that

$$f'(x) \approx \frac{f(x + \Delta x) - f(x)}{\Delta x}$$

as delta x approaches zero). A nested bisection routine should work as follows;

$$\frac{\delta \ln L(X)}{\delta \beta}$$

is to be driven to zero by bisection of interval $[\beta_{min}, \beta_{max}]$.

1. Given the current bisection interval $[\beta_L, \beta_R]$
2. Solve

$$\frac{\partial \ln L(X)}{\partial \theta} = 0$$

by bisection to get a θ for each shape parameter.

TABLE 2.11

StatGraphics Analysis, Left-Censored Weibull Data
200 values ranging from 0.2 to 1.1924
Number of left-censored observations: 33
Number of right-censored observations: 0

Fitted Distributions
Weibull
shape = 1.88297
scale = 0.476293

3. Calculate

$$\frac{\partial \ln L(X)}{\partial \beta}$$

for β_L and β_R and their corresponding scale parameters.

4. If the results are of opposite signs, the solution lies inside the current bisection interval. If they are of the same sign, the solution is outside the current bisection interval.

5. Iterate until convergence takes place.

Assume that the Weibull data from the simulation with $\beta = 2$, $\theta = 0.5$ (Weibull_b2_theta_0.5 in the simulated dataset) are parts per million of an impurity, and the instrument's lower detection limit is 0.20. StatGraphics returns the results in Table 2.11 and Figure 2.20.

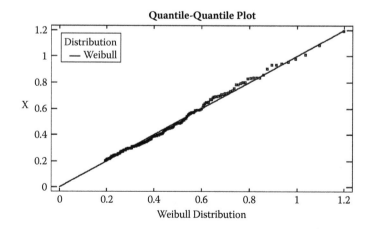

FIGURE 2.20
Quantile-Quantile plot, left-censored Weibull data.

Bisection fit, left-censored Weibull distribution $\text{ORIGIN} := 1$

$n := \text{rows}(x)$ $i := 1..n$ Detection limit $D := 0.20$

Number of data above the detection limit $r := \left[\sum_{i=1}^{n} x_i > D \right]$ $r = 167$

Log likelihood function: the conditional test $x > D$ excludes the left-censored observations

$$\ln L(\beta,\theta) := r \cdot \ln(\beta) - r \cdot \beta \cdot \ln(\theta) + \sum_{i=1}^{n} \left[(\beta-1) \cdot \ln(x_i) - \left(\frac{x_i}{\theta} \right)^{\beta} \right] \cdot (x_i > D) + (n-r) \cdot \ln \left[1 - \exp \left[-\left(\frac{D}{\theta} \right)^{\beta} \right] \right]$$

Function to be driven to zero $f(\beta,\theta) := \dfrac{d}{d\beta} \ln L(\beta,\theta)$

Function to get the scale parameter for any given shape $f\_theta(\beta,\theta) := \dfrac{d}{d\theta} \ln L(\beta,\theta)$

$$\text{Bisect\_theta}(\beta,c,d) := \begin{vmatrix} x_L \leftarrow c \\ x_R \leftarrow d \\ \text{while } x_R - x_L > 0.0001 \\ \quad \begin{vmatrix} \text{test} \leftarrow f\_theta(\beta,x_L) \cdot f\_theta(\beta,x_R) \\ \text{TEMP} \leftarrow x_L \ \text{if test} < 0 \\ x_R \leftarrow x_L \ \text{if test} > 0 \\ x_L \leftarrow \text{TEMP if test} > 0 \\ x_L \leftarrow \dfrac{x_L + x_R}{2} \end{vmatrix} \\ \text{Bisect} \leftarrow \dfrac{x_L + x_R}{2} \end{vmatrix}$$

FIGURE 2.21A
Nested bisection solution, left-censored Weibull data.

Figure 2.21 shows that these results can be replicated in MathCAD by means of the nested bisection loops discussed above. It was, however, necessary to manipulate the bisection interval for the scale parameter and also to increase the convergence tolerance from one millionth to 0.0001 to obtain convergence. Although it is relatively simple to find a working interval for a single bisection, adjustments to the interval for one part of a nested bisection may make the formerly valid interval for the other part unworkable. Computation time was noticeable as would be expected for a nested bisection solution. In practice, the lower detection limit will usually be so close to zero that treatment of the data as uncensored will provide very good starting estimates for the shape and scale parameters and, therefore, relatively tight bisection intervals.

Given the need to micromanage the bisection intervals, a response surface procedure such as evolutionary operation (EVOP) might be better. These

Range to bisect for beta $\begin{bmatrix} a \\ b \end{bmatrix} := \begin{bmatrix} 1.5 \\ 2 \end{bmatrix}$ Range to bisect for theta $\begin{bmatrix} c \\ d \end{bmatrix} := \begin{bmatrix} 0.2 \\ 4 \end{bmatrix}$

$\text{Bisect(a, b)} := \Bigg|$ $x_L \leftarrow a$ Make sure the solution is inside

$x_R \leftarrow b$ the interval, and that the bisection

while $x_R - x_L > 0.0001$ limits for theta will work for this interval.

$\quad\Bigg|$ $\theta_L \leftarrow \text{Bisect\_theta}(x_L, c, d)$ $\theta_L := \text{Bisect\_theta}(a, c, d)$

$\theta_R \leftarrow \text{Bisect\_theta}(x_R, c, d)$ $\theta_L = 0.453$

$\text{test} \leftarrow f(x_L, \theta_L) \cdot f(x_R, \theta_R)$

$\text{TEMP} \leftarrow x_L \quad \text{if test} < 0$ $\theta_R := \text{Bisect\_theta}(b, c, d)$

$x_R \leftarrow x_L \quad \text{if test} > 0$

$x_L \leftarrow \text{TEMP} \quad \text{if test} > 0$ $\theta_R = 0.483$

$x_L \leftarrow \dfrac{x_L + x_R}{2}$ $f(a, \theta_L) = 36.444$

$\text{Bisect} \leftarrow \dfrac{x_L + x_R}{2}$ $f(b, \theta_R) = -9.142$

$\beta := \text{Bisect(a, b)} \quad \theta := \text{Bisect\_theta}(\beta, c, d)$

$\begin{bmatrix} \beta \\ \theta \end{bmatrix} = \begin{bmatrix} 1.883 \\ 0.4763 \end{bmatrix}$

FIGURE 2.21B
Nested bisection solution, left-censored Weibull data, continued.

methods always work if (1) there is a unique maximum or minimum on the response surface, and (2) there are no local minima or maxima. Figure 2.22, a contour plot of the log likelihood function for combinations of 10 shape and 10 scale parameters, shows that this is indeed the case for the left-censored Weibull data.

Furthermore, the maximum is obviously between 1.8 and 2.0 for the shape parameter and in the interval [0.45,0.49] for the scale parameter. *This suggests an EVOP-like approach for optimization on a response surface.*

Distribution Fitting by Response Surface Methods

Given the need to optimize two parameters, create a box with corner points $[a_1, b_1]$ and $[a_2, b_2]$ with

$$\left(\frac{a_1 + b_1}{2}, \frac{a_2 + b_2}{2} \right)$$

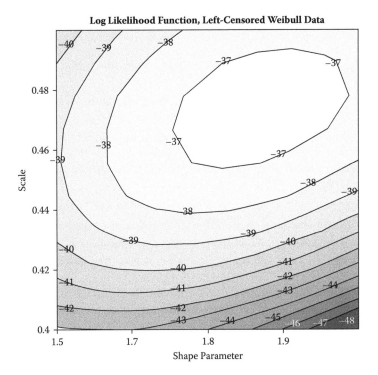

FIGURE 2.22
Log likelihood, left-censored Weibull data.

as the center point. Calculate the objective function, in this case the log likelihood function, for all five points. The algorithm is then extremely simple:

1. If a corner position contains the best result, it becomes the center point for the next iteration. This causes the box or rectangle to seek the top of the response surface. (Make sure none of the new corner points exceeds a practical limit, e.g., zero or less for the scale parameter.)

2. If the center position contains the best result, the optimum lies somewhere inside the rectangle. Contract the square by 50% on each side and repeat the calculations for the new corner squares.

3. Continue until the square contracts to the desired tolerance for the solution.

Visual Basic for Applications function FitCensored.bas performs this procedure automatically, and obtains $\beta = 1.8832$, $\theta = 0.47633$, and a maximum log likelihood of -36.531 when the tolerance is set to 0.0001. The format of the function is FitCensored(Data, Distribution, Lower_Detection, shape_min, shape_max, scale_min, scale_max, tolerance) where:

- Data is the range of cells on the spreadsheet that contain the measurements.
- Distribution is Weibull or Gamma. The function reads the first character and will understand lower or upper case.
- Lower_Detection is the lower detection limit, and it tells the function whether a measurement has been left-censored. If a measurement is *greater than* this number, it is treated as non-censored (toward the count of r).
- shape_min and shape_max define the axis of the starting box.
- scale_min and scale_max define the ordinates of the starting box.
- tolerance defines the stopping rule. The function stops when the change in both the shape and scale parameters is less than this number.

An advantage of this approach is that it is not mandatory, as it is for bisection, for the starting box to contain the answer. If the initial boundaries for the shape parameter are [1.6,1.8] and those for the scale parameter are [0.40,0.44], the initial rectangle does not contain the result [1.8832,0.47633]. The algorithm finds its way there nonetheless because the box "climbs" the response surface toward this maximum. If the initial boundaries are far from the answer, for example, [1.2,1.3] for beta and [0.2,0.25] for theta, the result is off slightly: [1.883,0.47627]. The log likelihood for this point is, however, −36.531 which is indistinguishable from that at [1.8832,0.47663]. Furthermore, the difference is in the ten-thousandths place, and is therefore unlikely to be of practical significance in a real factory application. The lesson here is that the algorithm is extremely robust.

The algorithm is even simple enough for deployment on an Excel spreadsheet as shown in Table 2.12A. In the first iteration, the center point delivers the highest log likelihood and therefore becomes the center point for the second iteration. The user must define r, the number of non-censored observations, manually although it can be computed easily enough by having the spreadsheet count the number of measurements that exceed the lower detection limits.

In the first iteration for Table 2.12, the user defines the dimensions of the rectangle in columns E and F.

C8 =(E8−E7)/2

C9 =(F8−F9)/2

G6={G$3*LN(E6)−G$3*E6*LN(F6)+SUM(((E6−1)*LN(A$6:A$205)−(A$6:A$205/F6)^E6)*(A$6:A$205>0.2))+(200−G$3)*LN(1−EXP(−((0.2/F6)^E6)))} where A6 through A205 hold the measurements and 0.2 is the lower detection limit. Note that this is an array summation. Cells G7 through G10 are similar, and they use the adjacent shape and scale parameters.

TABLE 2.12A

Response Surface Distribution Fitting (First Two Iterations)

	B	C	D	E	F	G
5	**Iteration**		**Point**	**Shape**	**Scale**	**log Likelihood**
6	1		**Center**	**1.90000**	**0.47000**	**−36.6289**
7			Upper left	1.80000	0.49000	−37.2919
8	DX	0.1000	Upper right	2.00000	0.49000	−37.1534
9	DY	0.0200	Lower left	1.80000	0.45000	−37.5346
10			Lower right	2.00000	0.45000	−39.1860
11	2		Center	1.90000	0.47000	−36.6289
12			Upper left	1.85000	0.48000	−36.6229
13	DX	0.0500	Upper right	1.95000	0.48000	−36.7109
14	DY	0.0100	Lower left	1.85000	0.46000	−36.9046
15			Lower right	1.95000	0.46000	−37.4336

The second and subsequent iterations determine which log likelihood is the largest. If that of the center point is the largest, DX and DY are divided in half. The center point is set to the coordinates of the point that had the largest log likelihood, and the corner points are then changed accordingly. Up to seven IF statements can be nested as shown, but only four are needed.

TABLE 2.12B

Response Surface Distribution Fitting (Final Iterations)

B	C	D	E	F	G
Iteration		**Point**	**Shape**	**Scale**	**log Likelihood**
19		Center	1.88320	0.47633	−36.530686
		Upper left	1.88301	0.47637	−36.530691
DX	0.0002	Upper right	1.88340	0.47637	−36.530694
DY	0.0000	**Lower left**	**1.88301**	**0.47629**	**−36.530684**
		Lower right	1.88340	0.47629	−36.530693
20		**Center**	**1.88301**	**0.47629**	**−36.530684**
		Upper left	1.88281	0.47633	−36.530687
DX	0.0002	Upper right	1.88320	0.47633	−36.530686
DY	0.0000	Lower left	1.88281	0.47625	−36.530687
		Lower right	1.88320	0.47625	−36.530692
21		**Center**	**1.88301**	**0.47629**	**−36.530684**
		Upper left	1.88291	0.47631	−36.530684
DX	0.0001	Upper right	1.88311	0.47631	−36.530685
DY	0.0000	Lower left	1.88291	0.47627	−36.530685
		Lower right	1.88311	0.47627	−36.530687

C13 =IF(G6=MAX(G6:G10),C8/2,C8)

C14 =IF(G6=MAX(G6:G10),C9/2,C9)

E11 =IF($G6=MAX($G6:$G10),E6,IF($G7=MAX($G6:$G10),E7,
    IF($G8=MAX($G6:$G10),E8,IF($G9=MAX($G6:$G10),E9,E10))))

F11 =IF($G6=MAX($G6:$G10),F6,IF($G7=MAX($G6:$G10),F7,
    IF($G8=MAX($G6:$G10),F8,IF($G9=MAX($G6:$G10),F9,F10))))

E12 =E11-C13 (left = center shape parameter minus DX)

F12 =F11+C14 (upper = center scale parameter plus DY)

The other corner positions are calculated similarly.

The algorithm converges on $b = 1.8830$ and $q = 0.47629$ as shown in the continuation of Table 2.12B. Note that DY drops below the stopping rule tolerance of 0.0001 that was used in the FitCensored function, but this is not a major issue. It was also necessary to format the log likelihoods to display to the millionths place to make the differences visible. The key point is that a seemingly complex algorithm can be performed on a spreadsheet, and then iterated by simply copying and pasting five row iterations until convergence occurs.

Figure 2.23 shows the iterative procedure graphically.

Minitab 15 also handles left-censored Weibull data, with the input in life table format as shown in Table 2.13 (be sure to select Maximum Likelihood as the estimation method to reproduce the results shown here). The interval starts and ends are identical when the exact measurements above the

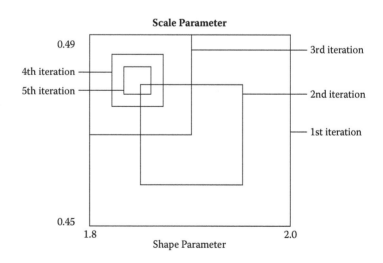

FIGURE 2.23
Response surface approach.

TABLE 2.13

Input Format for Left-Censored
Data, Minitab

Start	End	Count
*	0.2000	33
0.2031	0.2031	1
0.2039	0.2039	1
0.2132	0.2132	1
0.2135	0.2135	1
0.2148	0.2148	1
0.2182	0.2182	1
0.2190	0.2190	1
...	...	...

detection limit are known. The results appear in Table 2.14, and the program also creates a probability plot.

It is therefore indeed possible to obtain maximum likelihood estimates for the parameters of a Weibull distribution when the data are left-censored, so it is possible to handle quality characteristics with lower detection limits.

The gamma distribution applies to impurities far more frequently, but the previous example shows that a maximum likelihood estimate is possible for any distribution as long as an equation exists for the log likelihood function. StatGraphics can in fact handle the left-censored gamma distribution.

TABLE 2.14

Minitab Results for Left-Censored Weibull Data

Censoring Information	Count
Uncensored Value:	167
Interval Censored Value:	33

Estimation Method: Maximum Likelihood
Distribution: Weibull

Parameter Estimates

Parameter	Estimate	Standard Error	95.0% Normal CI Lower	Upper
Shape	1.8830	0.1108	1.6779	2.1131
Scale	0.47629	0.01904	0.44040	0.51511

Log-Likelihood = −36.529

Goodness-of-Fit
Anderson-Darling (adjusted) = 3.3140

Left-Censored Gamma Distribution

The probability density function of the two-parameter gamma distribution is as follows:

$$f(x) = \frac{\gamma^{\alpha}}{\Gamma(\alpha)} x^{\alpha-1} \exp(-\gamma x)$$

Then, given r out of n observations above detection limit D,

$$L(X) = \prod_{i=1}^{r} \frac{\gamma^{\alpha}}{\Gamma(\alpha)} x_i^{\alpha-1} \exp(-\gamma x_i) \times \prod_{j=1}^{n-r} \int_0^D \frac{\gamma^{\alpha}}{\Gamma(\alpha)} x^{\alpha-1} \exp(-\gamma x) dx$$

$$\ln L(X) = r\alpha \ln \gamma - r \ln(\Gamma(\alpha)) + (\alpha - 1) \sum_{i=1}^{r} \ln(x_i) - \gamma \sum_{i=1}^{r} x_i$$

$$+ (n-r) \ln \left(\int_0^D \frac{\gamma^{\alpha}}{\Gamma(\alpha)} x^{\alpha-1} \exp(-\gamma x) dx \right)$$

Consider the set of simulated data from the gamma distribution with $\alpha = 8$ and $\gamma = 2$ (Gamma_a8_g2 in the simulated data spreadsheet), and a lower detection limit of 2.5. Figure 2.24 shows the procedure in MathCAD, whose results match those from StatGraphics (Table 2.15), which also provides a quantile-quantile plot. It was necessary to experiment with the bisection limits in the MathCAD program to get convergence.

FitCensored.bas finds $\alpha = 7.7709$ and $\gamma = 1.8407$ given a starting rectangle of [6,9] for alpha, and [1.5,3] for gamma, and a tolerance of 0.00001.

This section has shown that it is feasible to fit the Weibull and gamma distributions to left-censored data, which means that control charts can be deployed for quality characteristics that have lower detection limits. It is also possible to get at least a point estimate for the nonconforming fraction and therefore the process performance index. The same approach works for other distributions, although it is unlikely that a measurement with a lower detection limit would follow a normal distribution.

This chapter has shown how to fit nonnormal distributions to process data, and how to develop charts for individual measurements and process averages. The next chapter will address the issue of charts for process variation.

Exercises

Exercise 2.1

The first chapter introduced a process that makes a chemical with an upper specification limit of 16 ppm for impurities. EXERCISES.XLS, tab 2-1, contains

$n := rows(x) \quad i := 1..n \quad \text{Detection limit} \quad D := 2.5$

Number of data above the detection limit $\quad r := \left[\displaystyle\sum_{i=1}^{n} x_i > D \right] \quad r = 89$

Log likelihood function: the conditional test

$x > D$ excludes the left-censored observations

$$lnL(\alpha, \gamma) := r \cdot \alpha \cdot \ln(\gamma) - r \cdot \ln(\Gamma(\alpha)) + \left[\sum_{i=1}^{n} [(\alpha - 1) \cdot \ln(x_i) \cdot (x_i > D) - \gamma \cdot x_i \cdot (x_i > D)] \right] \cdots$$

$$+ (n - r) \cdot \ln \left[\int_{0}^{D} \frac{\gamma^{\alpha}}{\Gamma(\alpha)} \cdot y^{\alpha - 1} \cdot \exp(-\gamma \cdot y) dy \right]$$

Function to be driven to zero $\quad f(\alpha, \gamma) := \dfrac{d}{d\alpha} lnL(\alpha, \gamma)$

Function to get the scale parameter for any given shape $\quad f\_gamma(\alpha, \gamma) := \dfrac{d}{d\gamma} lnL(\alpha, \gamma)$

$\text{Bisect\_gamma}(\alpha, c, d) := \begin{vmatrix} x_L \leftarrow c \\ x_R \leftarrow d \\ \text{while } x_R - x_L > 0.0001 \\ \quad \begin{vmatrix} \text{test} \leftarrow f\_gamma(\alpha, x_L) \cdot f\_gamma(\alpha, x_R) \\ \text{TEMP} \leftarrow x_L \quad \text{if test} < 0 \\ x_R \leftarrow x_L \quad \text{if test} > 0 \\ x_L \leftarrow \text{TEMP} \quad \text{if test} > 0 \\ x_L \leftarrow \dfrac{x_L + x_R}{2} \end{vmatrix} \\ \text{Bisect} \leftarrow \dfrac{x_L + x_R}{2} \end{vmatrix}$

FIGURE 2.24A
Nested bisection solution, left-censored gamma distribution.

50 samples of 10 measurements. Given the 500 individual measurements on the worksheet, determine the following:

1. The parameters of the three-parameter gamma distribution that are the best fit for the measurements
2. The process performance index *PPU*
3. The control limits and center line of the control chart for individuals
4. The control limits and center line of the chart for sample averages

Exercise 2.2

A process conforms to a Weibull distribution with a shape parameter $\beta = 2.0$, scale parameter $\theta = 1.5$, and threshold parameter $\delta = 3.0$.

$$\text{Range to bisect for alpha} \begin{bmatrix} a \\ b \end{bmatrix} := \begin{bmatrix} 6 \\ 9 \end{bmatrix} \qquad \text{Range to bisect for gamma} \begin{bmatrix} c \\ d \end{bmatrix} := \begin{bmatrix} 1.4 \\ 2.3 \end{bmatrix}$$

$$\text{Bisect (a,b)} := \begin{vmatrix} x_L \leftarrow a \\ x_R \leftarrow b \\ \text{while } x_R - x_L > 0.0001 \\ \begin{vmatrix} \gamma_L \leftarrow \text{Bisect\_gamma}(x_L, c, d) \\ \gamma_R \leftarrow \text{Bisect\_gamma}(x_R, c, d) \\ \text{test} \leftarrow f(x_L, \gamma_L) \cdot f(x_R, \gamma_R) \\ \text{TEMP} \leftarrow x_L \text{ if test<0} \\ x_R \leftarrow x_L \text{ if test>0} \\ x_L \leftarrow \text{TEMP if test>0} \\ x_L \leftarrow \dfrac{x_L + x_R}{2} \end{vmatrix} \\ \text{Bisect} \leftarrow \dfrac{x_L + x_R}{2} \end{vmatrix}$$

Make sure the solution is inside
the interval, and that the bisection
limits for theta will work for this interval.

$\gamma_L := \text{Bisect\_gamma}(a, c, d)$

$\gamma_L = 1.424$

$\gamma_R := \text{Bisect\_gamma}(b, c, d)$

$\gamma_R = 2.129$

$f(a, \gamma_L) = 1.653$

$f(b, \gamma_R) = -0.753$

$\alpha := \text{Bisect}(a, b) \qquad \gamma := \text{Bisect\_gamma}(\alpha, c, d)$

$$\begin{bmatrix} \alpha \\ \gamma \end{bmatrix} = \begin{bmatrix} 7.772 \\ 1.841 \end{bmatrix}$$

FIGURE 2.24B
Nested bisection solution, left-censored gamma, continued.

TABLE 2.15

StatGraphics Results, Left-Censored Gamma Distribution
100 values ranging from 2.5 to 9.113
Number of left-censored observations: 11
Number of right-censored observations: 0

Fitted Distributions
Gamma
shape = 7.77132
scale = 1.84081

1. What are the Shewhart-equivalent (0.00135 false alarm risks) and center line of the chart for individuals?
2. If it were desired to use the Western Electric zone tests, what are the upper and lower boundaries for Zone A (±2 sigma equivalent)?

Exercise 2.3

EXERCISES.XLS tab 2-3 contains 100 measurements (parts per million of a trace impurity) that were simulated from a gamma distribution with $\alpha = 2$ and $\gamma = 20$. The lower detection limit is 0.03 parts per million. Fit a gamma distribution to the censored data, and find the correct center line and 0.99835 quantile for the upper control limit. If the upper specification limit for the impurity is 0.5 ppm, what is the process performance index?

Solutions

Exercise 2.1

1. Determine the parameters of the distribution.

   The Visual Basic for Applications function FitGamma3.bas finds $\alpha = 2.632$, $\gamma = 0.8864$, and $\delta = 3.108$. StatGraphics Centurion returns similar results. The most convenient way to do this is with menu SPC: Capability Analysis, which allows simultaneous input of all 10 data columns. Describe: Distribution Fitting requires all 500 data to be in a single column.

   Distribution: Gamma (3-Parameter)

   sample size = 500

   average group size = 10.0

   shape = 2.63135

   scale = 0.886196

   threshold = 3.10806

   (mean = 6.07732)

   (sigma = 1.83046)

   StatGraphics also provides a histogram and quantile-quantile plot, both of which suggest that the model is valid. The practitioner should always expect to see these figures in any process capability report. The figure says the report is for columns 1 through 9, but column 10 was included (noting that the total data count is 500). If the program treats the column names as text, Col_10 would be between Col_1 and Col_2.

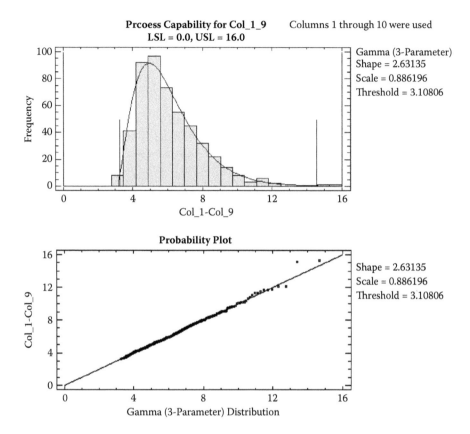

The user can also select a chi square goodness-of-fit test, which is acceptable for this data. Use of ChiSquare_Gamma yields (with a start of 30 cells) a chi square test statistic of 6.89 for 18 cells after combination of cells in the upper tail to meet the requirement that the expected count be at least 5.

2. Compute the process performance index.

StatGraphics will compute the continuous probability density function of a number when given the distribution parameters; this is available through Describe: Distribution Fitting: Probability Distributions. The upper tail area above 16 is 0.000456, for which the corresponding standard normal deviate divided by 3 is 1.10. The capability analysis returns the same result and quotes 456 defects per million opportunities.

=1–GAMMADIST(16–3.108,2.632,1/0.8864,1) in Microsoft Excel also returns 456 DPMO, and then=–NORMSINV(1–GAMMADIST(16–3.108,2.632,1/0.8864,1))/3 returns the process performance index of 1.10.

3. Compute the center line and control limits for individuals.

UCL =GAMMAINV(0.99865,2.632,1/0.8864)+3.108 = 14.58

CL =GAMMAINV(0.50,2.632,1/0.8864)+3.108 = 5.71

LCL =GAMMAINV(0.00135,2.632,1/0.8864)+3.108 = 3.27

StatGraphics returns the following when the 500 measurements are provided in one column to generate the chart for individuals. Two false alarms are more than expected (0.675) but not impossible.

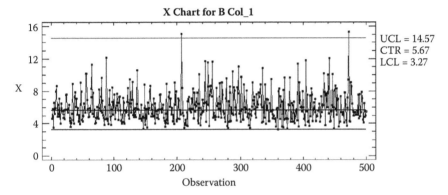

4. Determine the control limits and center line of the chart for sample averages.

The average of n measurements from a gamma distribution with parameters (α, γ, δ) follows another gamma distribution with parameters $(n\alpha, n\gamma, \delta)$. The average of 10 measurements from a gamma distribution with $\alpha = 2.632$, $\gamma = 0.8864$, and $\delta = 3.108$ should therefore follow one with $\alpha = 26.32$, $\gamma = 8.864$, and $\delta = 3.108$. The control limits are

UCL =GAMMAINV(0.99865,26.32,1/8.864)+3.108 = 8.12

CL =GAMMAINV(0.50,26.32,1/8.864)+3.108 = 6.04

LCL = GAMMAINV(0.00135,26.32,1/8.864)+3.108 = 4.64

The large shape parameter suggests that this will be almost indistinguishable from a normal distribution with mean $3.108 + 26.32/8.864 = 6.08$, variance $26.32/8.864^2 = 0.335$, and standard deviation 0.579. When the 50 sample averages are given to StatGraphics as individual measurements from a presumably normal distribution, they have an average of 6.08 and a standard deviation of 0.570. The averages pass the chi square for goodness of fit to the normal distribution, but the histogram and the failure of the points in the quantile-quantile plot to scatter randomly around the regression line (figure below) still suggest that the averages did not come from a normal distribution. This is why the quality practitioner should always insist on seeing the histogram and quantile-quantile plot for any process capability study or SPC chart setup.

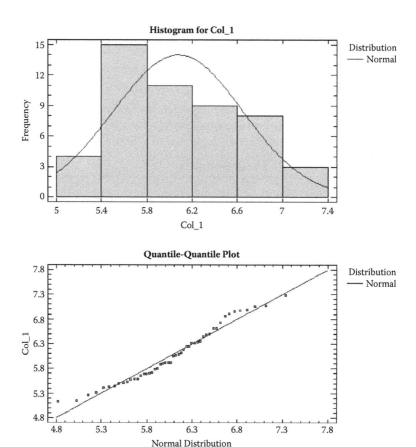

Exercise 2.2

1. Determine the Shewhart-equivalent control limits and center line for a Weibull distribution with a shape parameter $\beta = 2.0$, scale parameter $\theta = 1.5$, and threshold parameter $\delta = 3.0$. Use the equation

$$F^{-1}(q) = \theta(-\ln(1-q))^{1/\beta} + \delta$$

   UCL: $F^{-1}(0.99865) = 1.5(-\ln(1-0.99865))^{1/2} + 3 = 6.856$

   Similarly, CL = 4.249 and LCL = 3.055.

2. Determine the boundaries (±2 sigma) of the Western Electric Zone A test.

   ±2 sigma corresponds to the 0.02275 and 0.97725 quantiles of the normal distribution.

Upper zone A: $F^{-1}(0.97725) = 1.5(-\ln(1-0.97725))^{1/2} + 3 = 5.918$ and the lower Zone A boundary is similarly 3.228.

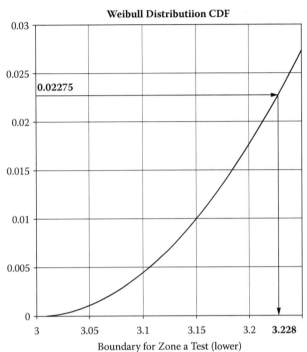

Weibull Distributiion CDF

Boundary for Zone a Test (lower)

Exercise 2.3

=FitCensored(K7:K106,"Gamma",0.03,2,4,10,20,0.0001) yields $\alpha = 2.0103$ and $\gamma = 19.919$ and a log likelihood of 101.02. The iterative spreadsheet algorithm that is provided with the exercise yields $\alpha = 2.011$ and $\gamma = 19.92$ (after stopping when the change in the scale parameter decreases to 0.0006) and a log likelihood of 101.02.

StatGraphics finds $\alpha = 2.0$ and $\gamma = 19.8181$, and it provides the following histogram and quantile-quantile plots. Remember to designate left-censored data with –1 in the column that holds the censoring information, and non-censored data with 0. The log likelihood with these parameters is 101.015. Any of these parameter sets should yield practical control limits and an accurate point estimate for the process performance index.

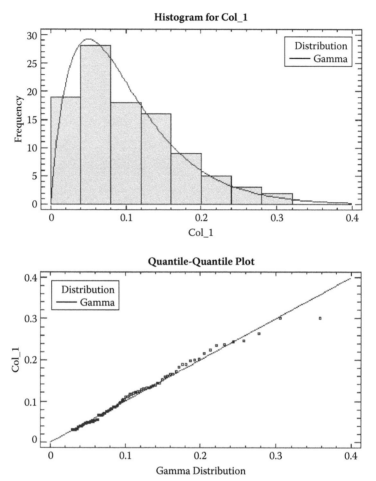

For $\alpha = 2.01$ and $\gamma = 19.92$, the center line is (in Microsoft Excel) =GAMMAINV(0.5,2.01,1/19.92) or 0.0848 and the upper control limit is =GAMMAINV(0.99865,2.01,1/19.92) or 0.448. The nonconforming fraction is =1–GAMMADIST(0.5,2.01,1/19.92,1) or 0.000528 and the corresponding process performance index is $-\frac{1}{3}\Phi^{-1}(0.000528)$ or 1.092

For $\alpha = 2.0$ and $\gamma = 19.82$, the center line and upper control limit are 0.0847 and 0.449 as calculated from Excel, and the process performance index is 1.089. StatGraphics obtains ("Critical values" table) 0.50 and 0.99865 quantiles of 0.0847 and 0.449 and a nonconforming fraction ("Tail Areas" table) of 0.0005413, which corresponds to PPU = 1.089.

Endnotes

1. The vertical line is shorthand for "given," e.g., "function of x *given* parameter set theta."
2. Minitab uses imported results from MathCAD to create this contour plot.
3. Page (1994) also discusses the Johnson transformation.
4. *NIST/SEMATECH e-Handbook of Statistical Methods* at http://www.itl.nist.gov/div898/handbook/eda/section3/eda35f.htm says, "The test statistic follows, approximately, a chi-square distribution with $(k - c)$ degrees of freedom where k is the number of non-empty cells and c = the number of estimated parameters (including location and scale parameters and shape parameters) for the distribution + 1. For example, for a 3-parameter Weibull distribution, $c = 4$."
5. See also Wasserman, 2000, for maximum likelihood estimation of Weibull metrics on a spreadsheet.
6. Per http://dspace.dsto.defence.gov.au/dspace/handle/1947/4548 as of 6/15/09, "A formula in closed form approaches *the distribution of the sum of Weibull distributed samples, which does not have close[d] form*, has been proposed."

3

Range Charts for Nonnormal Distributions

The central limit theorem says nothing about ranges and standard deviations from nonnormal distributions, and even the distribution of ranges from a normal distribution is nonnormal. The long upper tail of the gamma distribution suggests that its sample ranges and standard deviations will be higher than those from a normal distribution with the same variance.

It is necessary to understand up front what the nonnormal control charts will measure. The mean and variance of the normal distribution are independent, so the chart for averages reflects only the mean, and the chart for variation reflects only the variance. This is not true for the gamma and Weibull distributions, where the mean and variance are both functions of the distributions' shape and scale parameters. It is therefore reasonable to expect a process shift to manifest itself in some manner on both control charts. In the case of the exponential distribution, however, a chart for variation is likely to be redundant because the mean and variance are both functions of the distribution's single parameter.

This chapter will develop a practical range chart for nonnormal distributions, but the first step is to understand the traditional approach for the normal distribution.

Traditional Range Charts

The center line and control limits for the range chart are as follows when the population standard deviation σ is known:

$$LCL = D_1\sigma \quad CL = d_2\sigma \quad UCL = D_2\sigma$$

where $D_1 = d_2 - 3d_3$ and $D_2 = d_2 + 3d_3$ (ASTM, 1990, 91–94). This suggests that the expected range is $d_2\sigma$ while $d_3\sigma$ is the range's standard deviation. Wilks (1948, 21) confirms that $d_2\sigma$ is in fact the expected range:[1]

$$E(R) = \int_{-\infty}^{\infty} (1 - F(x)^n - (1 - F(x))^n)\,dx \tag{3.1}$$

and ASTM (1990, 92) provides a similar formula for d_2:

$$d_2 = \int_{-\infty}^{\infty} (1 - \Phi(x_1)^n - \left(1 - \Phi(x_1)\right)^n)\,dx$$

The resulting $\bar{R} \pm 3\sigma_R$ are only approximations even when the underlying distribution is normal (Montgomery, 1991, 226). Consider a sample of four. $D_2 = 4.698$ and, if applied to a standard normal distribution whose standard deviation is 1.0, the resulting false alarm risk is 0.00495 versus the expected 0.00135. It is easy to imagine that the situation for a skewed distribution such as the gamma distribution will be even worse. This suggests the desirability of R charts with known and exact false alarm risks, and the next section will develop them.

Range Charts with Exact Control Limits

Wilks (1948, 21) gives the following expression for the probability density function $h(R)$ of the range of any distribution $f(x)$. Its integration yields the cumulative density function $H(R)$ of the range.

$$h(R)dR = n(n-1)\int_{-\infty}^{\infty} (F(R+x_{(1)}) - F(x_{(1)}))^{n-2} f(R+x_{(1)}) f(x_{(1)})\, dx_{(1)} dR \quad (3.2)$$

Although the exponential distribution is of limited application in statistical process control, it can serve as a good illustrative example because the above integral has a closed form for its range.

Range Chart for the Exponential Distribution

The probability density function for the exponential distribution is $f(x) = \gamma \exp(-\gamma x)$ and its cumulative probability is $F(x) = 1 - \exp(-\gamma x)$. Now apply Equation 3.2:

$$h(R)dR = n(n-1)\int_{0}^{\infty} (1-\exp(-\gamma(R+x_{(1)})) - (1-\exp(-\gamma x_{(1)})))^{n-2}\gamma^2$$

$$\times \exp(-\gamma(R+x_{(1)}))\exp(-\gamma x_{(1)})dx_{(1)}dR$$

$$= n(n-1)\int_{0}^{\infty} (\exp(-\gamma x_{(1)})$$

$$- \exp(-\gamma(R+x_{(1)})))^{n-2}\gamma^2 \exp(-\gamma R)\exp(-\gamma x_{(1)})^2 dx_{(1)}dR$$

$$= n(n-1)\int_{0}^{\infty} (1-\exp(-\gamma R))^{n-2}\gamma^2 \exp(-\gamma R)\exp(-\gamma x_{(1)})^n dx_{(1)}dR$$

$$= (n-1)(1-\exp(-\gamma R))^{n-2}\gamma^2 \exp(-\gamma R)\exp(-\gamma x_{(1)})^n dR \Big|_{0}^{\infty}$$

$$= -(n-1)(1-\exp(-\gamma R))^{n-2}\gamma^2 \exp(-\gamma R)dR$$

The next step is to integrate for the cumulative distribution function of R.

$$u = \exp(-\gamma R) \Rightarrow du = -\gamma \exp(-\gamma R)dR$$

$$\text{and then } h(R)dR = -(n-1)\left(1-\exp(-\gamma R)\right)^{n-2}\gamma\left(-d\exp(-\gamma R)\right)$$

$$H(R) = \int_0^R (n-1)\left(1-\exp(-\gamma R)\right)^{n-2}\gamma\left(d\exp(-\gamma R)\right)$$

$$= \left(1-\exp(-\gamma R)\right)^{n-1}$$

Then, for the range of n measurements from an exponential distribution with scale parameter γ, and q, any desired quantile of the range,

$$H(R) = (1-\exp(-\gamma R))^{n-1} \text{ and } H^{-1}(q) = \frac{-\ln\left(1-q^{\frac{1}{n-1}}\right)}{\gamma} \qquad \text{(Set 3.3)}$$

Consider 400 samples of four measurements from an exponential distribution with scale parameter $\gamma = 2$ and mean $1/\gamma = 0.5$ (Exponential_g2 in the simulated data spreadsheet). The average of the 1600 points is $\gamma = 0.51553$, and therefore $\gamma = 1.940$. Figure 3.1 shows a quantile-quantile plot of the ordered ranges versus the $(i - 0.5)/400$ quantile

$$\frac{-\ln\left(1-\left(\frac{i-0.5}{400}\right)^{\frac{1}{3}}\right)}{1.940}.$$

The actual choice of γ affects only the quantile-quantile plot's slope (which should ideally be close to 1) and not the correlation.

The 0.00135, 0.5, and 0.999865 quantiles of this distribution are 0.0604, 0.814, and 3.973, respectively. Figure 3.2 shows the resulting range chart for the first 100 points.

This section has used the exponential distribution to illustrate the development of range charts for nonnormal distributions, and the next section will address the far more common normal distribution.

Exact Control Limits for the Range Chart, Normal Distribution

In the case of the normal distribution, the equation from Wilks becomes

$$h(R)dR = n(n-1)\int_{-\infty}^{\infty} \left(\Phi(R+x_{(1)})-\Phi(x_{(1)})\right)^{n-2} f(R+x_{(1)})f(x_{(1)})dx_{(1)}dR \quad (3.4)$$

$$\text{quantile}(F, n, y) := \frac{-\ln\left(1 - F^{\frac{1}{n-1}}\right)}{Y}$$ Quantile of F (axis of the Q-Q plot)

$$R := \text{sort}(R) \qquad m: = \text{rows}(R) \qquad i := 1..m \qquad q_i := \text{quantile}\left(\frac{i-0.5}{m}, 4, 1940\right)$$

$$\begin{bmatrix} b_1 \\ b_0 \\ \text{corr} \end{bmatrix} := \begin{bmatrix} \text{slope}(q,R) \\ \text{intercept}(q,R) \\ \text{corr}(q,R) \end{bmatrix} \qquad \begin{bmatrix} b_1 \\ b_0 \\ \text{corr} \end{bmatrix} = \begin{bmatrix} 0.9459 \\ 0.0409 \\ 0.9985 \end{bmatrix}$$

FIGURE 3.1
Q-Q plot, sample ranges from an exponential distribution.

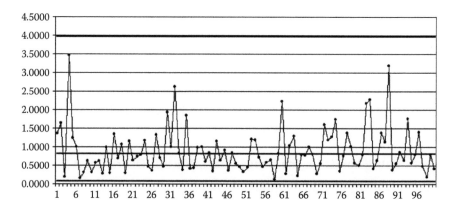

FIGURE 3.2
Control chart, sample ranges from an exponential distribution.

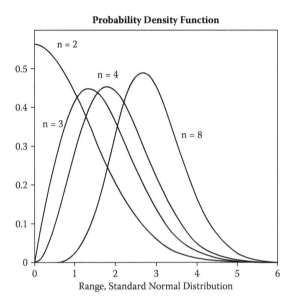

FIGURE 3.3
Probability density function, range of the normal distribution.

where $f(x)$ is the probability density function of the normal distribution. Figure 3.3 shows this probability density function for n = 2, 3, 4, and 8. Although it becomes more bell shaped with increasing sample size, it is clearly not a bell curve.

Beyer (1991, 270) tabulates percentiles of the range for samples that range from 2 to 20. The MathCAD routine in Figure 3.4 will reproduce these values (Table 3.1A), and it can also calculate the *exact* 0.00135 (lower control limit, corresponding to D_1), 0.5 (center line), and 0.99865 (upper control limit, corresponding to D_2) quantiles of the distribution of the range of a standard normal distribution. The resulting values appear in Table 3.1B; multiply them by the estimated standard deviation of the process to get the control limits and the center line. Note, however, that the user can set the false alarm risk to whatever is desirable by means of the algorithm shown below. Multiply the desired quantile by the standard deviation to set the control limits and center line.

These tables enable the user to set exact control limits, at least in applications where the tables provide the desired false alarm risks. These methods are otherwise of little use unless the practitioner can deploy them to the factory floor. Traditional range charts can of course be used, but remember that this material's primary purpose is for extension to nonnormal applications. The goal is therefore a spreadsheet function that will calculate the quantile of any range, which then allows a visual control, like a change in cell color, to tell the production

$$f(x) := \frac{1}{\sqrt{2\pi}} \exp\left(\frac{-1}{2} x^2\right) \quad \Phi(x) := \text{cnorm}(x) \text{ Standard normal pdf and cdf}$$

Probability density

function for the

range of a standard

$$h(R,n) := n \cdot (n-1) \cdot \int_{-9}^{+9} (\Phi(R + x_1) - \Phi(x_1))^{n-2} \cdot f(R + x_1) \cdot f(x_1) \, dx_1$$

normal deviation.

9 = "infinity"

for practical purposes.

$$H(R,n) := \int_0^{+9} h(R,n) \, dR \quad \text{cdf of the range}$$

Function to be driven to 0 by bisection:

$$g(R, n, q) := R(R, n) - q$$

qth quantile of the range for a sample of n

$$\text{Bisect}(n,q) := \begin{vmatrix} x_L \leftarrow 0 \\ x_R \leftarrow 7 \\ \text{while } x_R - x_L > 0.0001 \\ \quad \begin{vmatrix} \text{test} \leftarrow g(x_L, n, q) \cdot g(x_R, n, q) \\ \text{TEMP} \leftarrow x_L \text{ if test} < 0 \\ x_R \leftarrow x_L \text{ if test} > 0 \\ x_L \leftarrow \text{TEMP if test} > 0 \\ x_L \leftarrow \dfrac{x_L + x_R}{2} \end{vmatrix} \\ \text{Bisect} \leftarrow \dfrac{x_L + x_R}{2} \end{vmatrix}$$

FIGURE 3.4
Quantiles of the range of the standard normal distribution.

worker that the entry exceeds a control limit. As an example, if the desired one-sided false alarm risk is 0.5%, and $F(R)$ is the cumulative density function for the range, $F(R) < 0.005$ or $F(R) > 0.995$ would result in an out-of-control signal.

One approach relies on numerical integration by Simpson's Rule (Hornbeck, 1975, 150):

$$\int_a^b f(x) \, dx \approx \frac{\Delta x}{15} \begin{bmatrix} 7f(a) + 7f(b) + \Delta x \big(f'(a) + f'(b)\big) + \\ \cdots 14 \sum_{j=2,4,6\ldots}^{n-2} f(a + j\Delta x) + 16 \sum_{j=1,3,5\ldots}^{n-1} f(a + j\Delta x) \end{bmatrix} \quad (3.5)$$

TABLE 3.1A

Quantiles of the Range of the Standard Normal Distribution

n	0.001	0.005	0.01	0.025	0.05	0.1	0.5	0.9	0.95	0.975	0.99	0.995	0.999
2	0.002	0.009	0.018	0.044	0.089	0.178	0.954	2.326	2.772	3.170	3.643	3.970	4.654
3	0.060	0.135	0.191	0.303	0.431	0.618	1.588	2.902	3.315	3.682	4.120	4.424	5.064
4	0.199	0.343	0.434	0.595	0.760	0.979	1.978	3.240	3.633	3.984	4.403	4.694	5.309
5	0.367	0.555	0.665	0.850	1.030	1.261	2.257	3.479	3.858	4.197	4.603	4.886	5.484
6	0.535	0.749	0.87	1.066	1.253	1.488	2.472	3.661	4.030	4.361	4.757	5.034	5.619
7	0.691	0.922	1.048	1.251	1.440	1.676	2.645	3.808	4.170	4.494	4.882	5.154	5.730
8	0.835	1.075	1.205	1.410	1.600	1.836	2.791	3.931	4.286	4.605	4.987	5.255	5.823
9	0.966	1.212	1.343	1.550	1.740	1.973	2.915	4.037	4.387	4.700	5.078	5.342	5.903
10	1.085	1.335	1.467	1.673	1.863	2.095	3.024	4.129	4.474	4.784	5.157	5.418	5.973
11	1.193	1.446	1.578	1.784	1.973	2.202	3.121	4.211	4.552	4.859	5.227	5.485	6.036
12	1.293	1.547	1.679	1.884	2.071	2.299	3.207	4.285	4.622	4.925	5.290	5.546	6.093
13	1.385	1.639	1.771	1.976	2.161	2.387	3.285	4.351	4.685	4.985	5.348	5.602	6.144
14	1.470	1.724	1.856	2.059	2.243	2.467	3.356	4.412	4.743	5.041	5.400	5.652	6.191
15	1.549	1.803	1.934	2.136	2.319	2.541	3.422	4.468	4.796	5.092	5.448	5.699	6.234
16	1.623	1.876	2.007	2.207	2.389	2.609	3.482	4.519	4.845	5.139	5.493	5.742	6.274
17	1.692	1.944	2.074	2.274	2.454	2.673	3.538	4.568	4.891	5.183	5.535	5.783	6.311
18	1.757	2.008	2.137	2.336	2.515	2.732	3.591	4.613	4.934	5.224	5.574	5.820	6.346
19	1.818	2.068	2.197	2.394	2.572	2.788	3.640	4.655	4.974	5.262	5.611	5.856	6.379
20	1.875	2.125	2.253	2.449	2.626	2.840	3.686	4.694	5.012	5.299	5.645	5.889	6.412

TABLE 3.1B

Shewhart-Equivalent Limits, Range Chart for the Normal Distribution

n	0.00135 LCL	0.50 CL	0.99865 UCL
2	0.002430	0.954	4.533
3	0.070	1.588	4.950
4	0.221	1.978	5.200
5	0.397	2.257	5.377
6	0.569	2.472	5.515
7	0.729	2.645	5.627
8	0.874	2.791	5.721
9	1.006	2.916	5.803
10	1.126	3.024	5.874
11	1.236	3.121	5.938
12	1.336	3.207	5.995
13	1.428	3.285	6.047
14	1.513	3.356	6.094
15	1.592	3.422	6.138
16	1.666	3.482	6.179
17	1.735	3.538	6.217
18	1.800	3.591	6.252
19	1.860	3.640	6.286
20	1.918	3.686	6.318

Multiply by the standard deviation to set the control limits and center line.

This required several seconds for a total of 100,000 iterations: 1000 in the first integral and 100 in the second. This was adequate to reproduce the values at the top of Table 3.1 to the nearest thousandth, and usually, but not always, the nearest ten-thousandth, when the tabulated values are used as arguments for the function. (Note that the ranges themselves are given only to the nearest thousandth.) The calculation-intensive nature of Simpson's Rule plus the need to specify enough intervals to obtain accurate results suggest that something better is desirable. Romberg integration is an iterative procedure whose stopping rule requires the difference between the last and second-to-last results to be less than a specified tolerance.

Romberg Integration

Romberg integration (Hornbeck, 1975, 150–154) involves a series of trapezoidal rule estimates of the integral. For the mth iteration,

$$T_{1,m} = \frac{\Delta x}{2}\left[f(a) + f(b) + 2 \sum_{j=1}^{(2^\wedge m)-1} f\left(a + \frac{b-a}{2^{m-1}} \right) \right] \qquad (3.6a)$$

For the first iteration, this becomes simply $\frac{b-a}{2}(f(a)+f(b))$: the area of the trapezoid with base $(b-a)$ and unequal sides $f(a)$ and $f(b)$. A recursive algorithm makes the successive computations relatively easy:

$$T_{1,m} = \frac{1}{2}T_{m-1} + \frac{b-a}{2^{m-1}} \sum_{i=1}^{2^{\wedge}(m-2)} f\left(a+(2i-1)\frac{b-a}{2^{m-1}}\right) \tag{3.6b}$$

For example,

$$T_{1,3} = \frac{1}{2}T_{1,2} + \frac{b-a}{4}\left[f\left(a+\frac{b-a}{4}\right)+f\left(a+3\frac{b-a}{4}\right)\right]$$

$$T_{1,4} = \frac{1}{2}T_{1,3} + \frac{b-a}{8}\left[f\left(a+\frac{b-a}{8}\right)+f\left(a+3\frac{b-a}{8}\right)\right.$$

$$\left.+f\left(a+5\frac{b-a}{8}\right)+f\left(a+7\frac{b-a}{8}\right)\right]$$

The above procedure effectively divides the region to be integrated into quarters, eights, sixteenths, and so on. The recursive procedure avoids the need to re-compute terms that have already been calculated, and this is especially helpful when they are numerical integrals themselves. The next step is extrapolation of $T_{l,k}$ for $l=1$ to m and $k=1$ to $m-1$.

$$T_{l,k} = \frac{1}{4^{l-1}-1}\left(4^{l-1}T_{l-1,k+1}-T_{l-1,k}\right) \tag{3.6c}$$

The stopping rule can be

$$\left|T_{l,1}-T_{l-1,i}\right|<\varepsilon \text{ or } \left|\frac{T_{l,1}-T_{l-1,i}}{T_{l,i}}\right|<\varepsilon$$

where the second option is a relative convergence criterion. In the latter case, it is necessary to make sure that $T_{l,i}$ is not zero, which is possible on the first iteration if $f(a)$ and $f(b)$ are both zero or close enough to be rounded off to zero. It should not be possible for subsequent iterations because $f(a + (b - a)/2)$ is in the center of the range to be integrated, and the following algorithm performs four iterations before it attempts to test for convergence.

The Visual Basic for Applications (VBA) function CDF_Range_Normal.bas uses Romberg integration to find the cumulative distribution of the range of the standard normal deviation. CDF_Range_Normal.bas could conceivably be used in a spreadsheet to compute the cumulative distribution of any standardized range

$$\frac{R}{\sigma} = \frac{x_{max}}{\sigma} - \frac{x_{min}}{\sigma}$$

from normally distributed data, and to change the cell color if the result exceeds the desired false alarm limit.

The standard deviation (s) chart is actually superior to the range (R) chart because it uses the entire sample instead of only its maximum and minimum. Furthermore, it was shown previously that exact control limits can be defined for the sample standard deviations, and for any desired false alarm risk α. If $\chi^2 = (n-1)s^2/\sigma_0^2$ where s^2 is the sample's variance and σ_0^2 is the nominal or expected variance, the control limits for χ^2 are the $(\alpha/2)$ and $(1 - \alpha/2)$ quantiles of the chi square distribution with $n - 1$ degrees of freedom. Microsoft Excel's CHIINV function can provide these quantiles directly.

Average Run Length, Range Chart

The power of the range chart to detect a change in process standard deviation is $\gamma(\sigma,n) = 1 - F(UCL|\sigma,n)$, and the average run length is its reciprocal. Figure 3.5 shows average run lengths (ARLs) for sample sizes of 2, 4, and 9 for the Shewhart-equivalent upper control limit, the 0.99865 quantile of the range for the indicated sample sizes.

Development of the ARL for the s chart is considerably simpler, and it allows comparison of the powers of the two charts.

Average Run Length, s Chart

Noting that

$$\chi^2 = \frac{(n-1)s^2}{\sigma_0^2},$$

it is possible to set an exact upper control limit for any desired false alarm risk α as follows:

$$UCL^2 = \frac{\sigma_0^2}{n-1}\chi_\alpha^2$$

Excel function CHIINV$(1-q,n-1)$ returns χ^2 for the qth quantile [e.g., CHIINV(0.05,3) returns the 0.95 quantile for the distribution with 3 degrees of freedom]. Table 3.2 shows the Shewhart-equivalent control chart factors for $\sigma_0 = 1$.

$$f(x,\sigma) := \frac{1}{\sqrt{2 \cdot \pi} \cdot \sigma} \cdot \exp\left(\frac{-1}{2\sigma^2} \cdot x^2\right) \quad \Phi(x,\sigma) := \text{cnorm}\left(\frac{x}{\sigma}\right) \quad \text{Standard normal pdf and cdf}$$

$$h(R, n, \sigma) := n \cdot (n-1) \cdot \int_{-9}^{19} \left(\Phi(R + x_1, \sigma) - \Phi(x_1, \sigma)\right)^{n-2} \cdot f(R + x_1, \sigma) \cdot f(x_1, \sigma) dx_1$$

$$H(R, n, \sigma) := \int_{0}^{+R} h(R, n, \sigma) dR \quad \text{cdf of the range} \quad \text{ARL(UCL}, n, \sigma) := \frac{1}{1 - H(\text{UCL}, n, \sigma)}$$

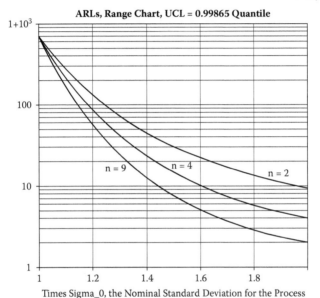

FIGURE 3.5
Average run lengths, range chart with UCL = 0.99865 quantile.

The Type II risk that s will be less than the *UCL* is

$$\beta = F\left(\frac{(n-1)UCL^2}{\sigma^2}\right)$$

where *F* is the cumulative distribution function for the chi square statistic with $n - 1$ degrees of freedom and *UCL* is the upper control limit for the s^2 chart. Figure 3.6 shows the average run lengths for sample sizes of 2, 4, and 9; MathCAD function pchisq(*x,v*) returns the cumulative chi square

TABLE 3.2

Exact Control Chart Factors for s and s² Charts

Sample	s² Chart (multiply by σ_0^2)			s Chart (Multiply by s_0)		
	LCL²	CL²	UCL²	LCL	CL	UCL
n	0.00135	0.5	0.99865	0.00135	0.5	0.99865
2	0.000	0.455	10.273	0.002	0.674	3.205
3	0.001	0.693	6.608	0.037	0.833	2.571
4	0.010	0.789	5.210	0.100	0.888	2.283
5	0.026	0.839	4.450	0.163	0.916	2.109
6	0.048	0.870	3.964	0.218	0.933	1.991
7	0.071	0.891	3.623	0.266	0.944	1.903
8	0.094	0.907	3.369	0.306	0.952	1.835
9	0.116	0.918	3.170	0.341	0.958	1.780
10	0.138	0.927	3.010	0.371	0.963	1.735
11	0.158	0.934	2.878	0.398	0.967	1.697
12	0.178	0.940	2.767	0.422	0.970	1.664
13	0.196	0.945	2.672	0.443	0.972	1.635
14	0.213	0.949	2.590	0.461	0.974	1.609
15	0.229	0.953	2.518	0.479	0.976	1.587
16	0.244	0.956	2.454	0.494	0.978	1.566
17	0.258	0.959	2.397	0.508	0.979	1.548
18	0.272	0.961	2.345	0.522	0.980	1.531
19	0.285	0.963	2.299	0.534	0.981	1.516
20	0.297	0.965	2.256	0.545	0.982	1.502

distribution for x with v degrees of freedom. The ARLs are only slightly lower than those of the corresponding ones for the R chart, which reinforces the perception that the R chart was "good enough" for the days in which the s chart was computationally prohibitive. The far greater problem consists of the fact that the normal approximation upon which the traditional R and s chart are both based is a very poor one, especially for small and even moderate sample sizes. It is hoped that this section has provided a practical and superior alternative.

Range Charts for Nonnormal Distributions

Development of cumulative density functions for the ranges of the exponential and the normal distributions suggests that the same can be done for gamma and Weibull distributions. The major barrier is the double integral, which becomes a triple integral if it is necessary to integrate the cumulative density function of the distribution itself. This is not a problem for the cumulative Weibull distribution, but the integral of the gamma distribution does not have a closed form for non-integral shape parameters. It is therefore best

$$ARL(UCL, n, \sigma) := \dfrac{1}{1 - pchisq\left[\dfrac{(n-1) \cdot UCL}{\sigma^2}, n-1\right]}$$

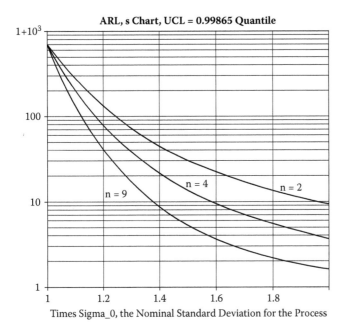

ARL, s Chart, UCL = 0.99865 Quantile

Times Sigma_0, the Nominal Standard Deviation for the Process

FIGURE 3.6
ARLs for the s chart with UCL = 0.99865 quantile.

to test assumptions on a gamma distribution whose shape parameter is an integer, and then continue to the general case.

Range Chart for the Gamma Distribution

The two-parameter gamma distribution is as follows, and the threshold parameter is not relevant to the range because subtraction of the smallest measurement from the largest cancels it out.

$$f(x) = \frac{\gamma^\alpha}{\Gamma(\alpha)} x^{\alpha-1} \exp(-\gamma x)$$

The cumulative distribution has a closed form only if α is an integer, noting that

$$\int x^m e^{ax} \, dx = \frac{x^m e^{ax}}{a} - \frac{m}{a} \int x^{m-1} e^{ax} \, dx$$

Consider the example of a gamma distribution with $\alpha = 3$ and $\alpha = 2$. Then

$$F(x) = \frac{\gamma^3}{\Gamma(3)} \int x^2 \exp(-\gamma x) dx$$

$$= \frac{\gamma^3}{\Gamma(3)} \left[\frac{-x^2 \exp(-\gamma x)}{\gamma} + \frac{2}{\gamma} \int x \exp(-\gamma x) dx \right]$$

$$= \frac{\gamma^3}{\Gamma(3)} \left[\frac{-x^2 \exp(-\gamma x)}{\gamma} + \frac{2}{\gamma} \left(\frac{-x \exp(-\gamma x)}{\gamma} + \frac{1}{\gamma} \int \exp(-\gamma x) dx \right) \right]$$

$$= \frac{\gamma^3}{\Gamma(3)} \left[\frac{-x^2 \exp(-\gamma x)}{\gamma} + \frac{2}{\gamma} \left(\frac{-x \exp(-\gamma x)}{\gamma} - \frac{1}{\gamma^2} \exp(-\gamma x) \right) \right]$$

$$= \frac{\exp(-\gamma x)}{\Gamma(3)} (-\gamma^2 x^2 - 2\gamma x - 2)$$

$$\frac{\exp(-\gamma x)}{\Gamma(3)} (-\gamma^2 x^2 + 2\gamma x - 2) \Big|_0^{\lvert x} = \frac{1}{2}((-\gamma^2 x^2 - 2\gamma x - 2)\exp(-\gamma x) + 2)$$

$$\Rightarrow F(x) = 1 - \frac{1}{2}(\gamma^2 x^2 + 2\gamma x + 2)\exp(-\gamma x)$$

Then the equation for the probability density function of the range becomes

$$h(R) = n(n-1) \int_0^\infty (F(R + x_{(1)}) - F(x_{(1)}))^{n-2} f(R + x_{(1)}) f(x_{(1)}) dx_{(1)}$$

and this is used in turn in the expression for the cumulative density.

One hundred samples of four were simulated from a gamma distribution with $\alpha = 3$, $\beta = 2$ (Gamma_a3_g2 in the simulated data spreadsheet). Figure 3.7 shows a MathCAD bisection routine for the $(i - 0.5)/n$ quantile of the range of a gamma distribution with the same parameters and a sample size of four; the bisection of 100 double integrals required more than 5 minutes. The resulting plot of the sorted ranges versus the quantiles yields a slope of 0.9199, an intercept of 0.1262, and a correlation coefficient of 0.9962. The slope of a Q-Q plot should be close to 1, and the actual results depend on the random seed for the dataset.

The VBA function CDF_Range_Gamma is similar to CDF_Range_Normal, and it can reproduce the cumulative distribution functions (CDFs) of the $(i - 0.5)/100$ quantiles of this gamma distribution to the nearest thousandth. (Note that those quantiles were themselves originally from a MathCAD bisection routine whose stopping rule is $(x_R - x_L) \leq 0.0001$.) The key difference is the upper limit of the second integral, for which $\infty = 10$ seemed adequate in this

$$y := 2 \quad f(x) := \frac{y^3}{\Gamma(3)} \cdot x^2 \cdot \exp(-y \cdot x) \quad F(x) := 1 - \frac{1}{2}(y^2 \cdot x^2 + 2y \cdot x + 2) \cdot \exp(-y \cdot x)$$

$$h(R, n) := n \cdot (n-1) \cdot \int_{+0}^{+10} (F(R + x_1) - F(x_1))^{n-2} \cdot f(R + x_1) \cdot f(x_1) dx_1$$

$$H(R, n) := \int_0^{+R} h(R, n) dR \quad \text{c df of the range}$$

Function to be driven to 0 by bisection:

$$g(R, n, q) := H(R, n) - q$$

qth quantile of the range for a sample of n

$$\text{Bisect } (n, q) := \begin{vmatrix} x_L \leftarrow 0 \\ x_R \leftarrow 5 \\ \text{while } x_R - x_L > 0.0001 \\ \quad \begin{vmatrix} \text{test} \leftarrow g(x_L, n, q) \cdot g(x_R, n, q) \\ \text{TEMP} \leftarrow x_L \text{ if test} < 0 \\ x_R \leftarrow x_L \text{ if test} > 0 \\ x_L \leftarrow \text{TEMP if test} > 0 \\ x_L \leftarrow \dfrac{x_L + x_R}{2} \end{vmatrix} \\ \text{Bisect} \leftarrow \dfrac{x_L + x_R}{2} \end{vmatrix}$$

$$R := \text{sort}(R) \quad n := \text{rows}(R) \quad I := 1..n \quad \text{axis}_i := \text{Bisect}\left(4, \frac{i - 0.5}{n}\right)$$

FIGURE 3.7A
Quantile-quantile plot, ranges from a gamma distribution.

case. It was necessary to increase it, though, to get accurate results for a higher range and a lower scale parameter. The user should, for any given application, try the high end of the ranges under consideration and make sure that the result of the integration does not change with this integration limit.

The next step would be to identify the 0.00135 (LCL), 0.5 (center line), and 0.99865 (UCL) quantiles of this distribution and then deploy them to a control chart. CDF_Range_Gamma could alternatively calculate the continuous distribution of any sample range and change the cell color if it was outside the desired false alarm limits.

A foreseeable practical problem involves the possible combinations of α and β, and the simplicity of the standardized normal distribution comes to mind. In the latter case, multiplication of the standardized range quantiles by the standard deviation yields appropriate control limits for any application. The relationship between the gamma distribution and the chi square distribution suggests a similar approach here.

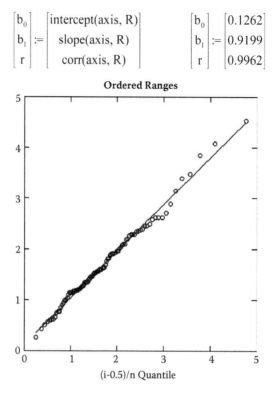

$$\begin{bmatrix} b_0 \\ b_1 \\ r \end{bmatrix} := \begin{bmatrix} \text{intercept(axis, R)} \\ \text{slope(axis, R)} \\ \text{corr(axis, R)} \end{bmatrix} \qquad \begin{bmatrix} b_0 \\ b_1 \\ r \end{bmatrix} := \begin{bmatrix} 0.1262 \\ 0.9199 \\ 0.9962 \end{bmatrix}$$

Ordered Ranges

(i-0.5)/n Quantile

FIGURE 3.7B
Quantile-quantile plot, ranges from a gamma distribution.

Range Chart for the Chi Square Distribution

It was shown in Chapter 2 that, if

$$f(x) = \frac{\gamma^\alpha}{\Gamma(\alpha)} x^{\alpha-1} \exp(-\gamma x),$$

$y = 2\gamma x$ follows a chi square distribution with $v = 2\alpha$ degrees of freedom. Microsoft Excel's CHIDIST function will not, however, perform the calculation for fractional degrees of freedom, so practical application will require the corresponding gamma distribution with $v = 2\alpha$ and $\gamma = 0.5$.

$$g(y) = \frac{1}{2^{\frac{v}{2}}\Gamma\left(\frac{v}{2}\right)} y^{\frac{v}{2}-1} \exp\left(-\frac{y}{2}\right) = \frac{\left(\frac{1}{2}\right)^{\frac{v}{2}}}{\Gamma\left(\frac{v}{2}\right)} y^{\frac{v}{2}-1} \exp\left(-\frac{y}{2}\right)$$

The idea is to deal with only one scale parameter so the range quantiles depend only on the shape parameter. In the case of the gamma distribution with $\alpha = 3$ and

$\gamma = 2$, $y = 4x$ should follow a chi square distribution with 6 degrees of freedom or a gamma distribution with $\alpha = 3$, $\gamma = 0.5$.

Consider the example of 2.500, the 0.835 quantile of the range of the gamma distribution with $\alpha = 3$, $\gamma = 2$, and a sample size of 4. The CDF of 4×2.500 for the range of the chi square distribution with 6 degrees of freedom, or alternatively the gamma distribution with $\alpha = 3$, $\gamma = 0.5$ should be 0.835, and the function CDF_Range_Gamma(10,4,3,0.5) gets 0.8350 after roundoff. The upper range of the nested integral must, however, be increased from 10 to 15.

Range Chart for the Weibull Distribution

The probability and cumulative density functions for the two-parameter Weibull distribution (the threshold parameter, if any, cancels out in the range calculation) are as follows. Although Excel contains no built-in function for this distribution, the cumulative distribution has a closed form so the calculations are no more intensive than those for the normal and gamma distributions.

$$f(x) = \frac{\beta}{\theta}\left(\frac{x}{\theta}\right)^{\beta-1}\exp\left(-\left(\frac{x}{\theta}\right)^{\beta}\right)$$

$$F(x) = 1 - \exp\left(-\left(\frac{x}{\theta}\right)^{\beta}\right)$$

The probability density function of the range becomes

$$h(R) = n(n-1)\int_0^{\infty}\left(-\exp\left(-\left(\frac{R+x_{(1)}}{\theta}\right)^{\beta}\right) + \exp\left(-\left(\frac{x_{(1)}}{\theta}\right)^{\beta}\right)\right)^{n-2} f(R+x_{(1)})f(x_{(1)})dx_{(1)}$$

$$= n(n-1)\int_0^{\infty}\left(\exp\left(-\left(\frac{x_{(1)}}{\theta}\right)^{\beta}\right) - \exp\left(-\left(\frac{R+x_{(1)}}{\theta}\right)^{\beta}\right)\right)^{n-2}\frac{\beta^2}{\theta^2}\left(\frac{R+x_{(1)}}{\theta}\right)^{\beta-1}\left(\frac{x_{(1)}}{\theta}\right)^{\beta-1}\cdots$$

$$\times\exp\left(-\left(\frac{R+x_{(1)}}{\theta}\right)^{\beta} - \left(\frac{x_{(1)}}{\theta}\right)^{\beta}\right)dx_{(1)}$$

$$h_{Weibull}(R) = n(n-1)\int_0^{\infty}\left(\exp\left(-\left(\frac{x_{(1)}}{\theta}\right)^{\beta}\right) - \exp\left(-\left(\frac{R+x_{(1)}}{\theta}\right)^{\beta}\right)\right)^{n-2}\cdots$$

$$\times\frac{\beta^2}{\theta^{2\beta}}((R+x_{(1)})x_{(1)})^{\beta-1}\exp\left(-\left(\frac{R+x_{(1)}}{\theta}\right)^{\beta} - \left(\frac{x_{(1)}}{\theta}\right)^{\beta}\right)dx_{(1)} \quad (3.7)$$

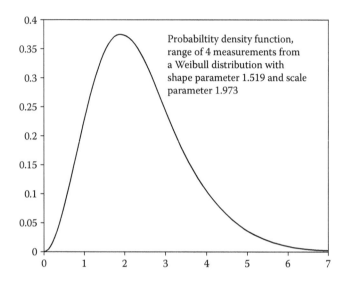

FIGURE 3.8
Distribution of sample ranges ($n = 4$) from a Weibull distribution.

One hundred samples of four were generated from a Weibull distribution with $\beta = 1.5$, $\theta = 2$, and $\delta = 1$ (Weibull_b1.5_theta_2_delta_1 in the simulated data spreadsheet). The VBA function FitWeibull3.bas yields a best fit of $\beta = 1.519$, $\theta = 1.973$, and $\delta = 1.023$, as does StatGraphics. Figure 3.8 shows the probability density function for the range of samples of 4 from this distribution.

Figure 3.9 illustrates the calculation of the $(i - 0.5)/100$ quantiles of the range and the resulting quantile-quantile plot with the sorted sample ranges on the ordinate.

The function CDF_Range_Weibull will return the cumulative distribution of R given the sample size, beta, and theta. As with CDF_Range_Gamma, it may be necessary to change the limit of the nested integral depending on the application.

Exercises

Exercise 3.1

The traditional upper control limit for the range of 3 measurements when the process standard deviation is known is $D_2\sigma$ where $D_2 = 4.358$. What is the actual false alarm risk for this control limit?

$$\text{TOL} := 0.0001 \quad \text{ORIGIN} := 1 \quad \begin{bmatrix} \beta \\ \theta \end{bmatrix} := \begin{bmatrix} 1.519 \\ 1.973 \end{bmatrix} \quad n := \text{rows}(R) \ n = 100$$

$$h(R,n) := n \cdot (n-1) \int_{+0}^{+10} \left[\exp\left[-\left(\frac{x_1}{\theta} \right)^{\beta} \right] - \exp\left[-\left(\frac{R+x_1}{\theta} \right)^{\beta} \right] \right]^{n-2} \cdot \frac{\beta^2}{\theta^{2\beta}} \cdot [(R+x_1) \cdot x_1]^{\beta-1}$$

$$\cdot \exp\left[-\left(\frac{R+x_1}{\theta} \right)^{\beta} - \left(\frac{x_1}{\theta} \right)^{\beta} \right] dx_1$$

$$H(R,n) := \int_0^{+R} h(R,n) \, dR \quad \text{cdf of the range}$$

Function to be driven to 0 by bisection:

$$g(R, n, q) := R(R, n) - q$$

qth quantile of the range for a sample of n

$$\text{Bisect}(n,q) := \begin{vmatrix} x_L \leftarrow 0 \\ x_R \leftarrow 8 \\ \text{while } x_R - x_L > 0.0001 \\ \quad \begin{vmatrix} \text{test} \leftarrow g(x_L, n, q) \cdot g(x_R, n, q) \\ \text{TEMP} \leftarrow x_L \text{ if test} < 0 \\ x_R \leftarrow x_L \text{ if test} > 0 \\ x_L \leftarrow \text{TEMP if test} > 0 \\ x_L \leftarrow \dfrac{x_L + x_R}{2} \end{vmatrix} \\ \text{Bisect} \leftarrow \dfrac{x_L + x_R}{2} \end{vmatrix}$$

$$R := \text{sort}(R) \quad n := \text{rows}(R) \quad I := 1..n \quad \text{axis}_i := \text{Bisect}\left(4, \frac{i-0.5}{n} \right)^i$$

FIGURE 3.9A
Quantile-quantile plot, ranges from a Weibull distribution.

Exercise 3.2

The traditional upper control limit for the standard deviation of 5 measurements when the process standard deviation is known is $B_6\sigma$ where $B_6 = 1.964$.

1. What is the actual false alarm risk for this control limit?
2. What are this control limit's power and average run length if the standard deviation increases 27.5%?

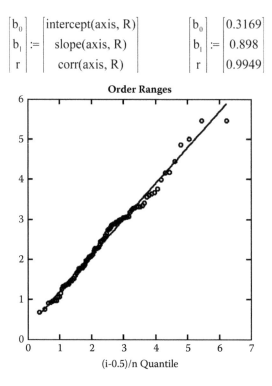

$$\begin{bmatrix} b_0 \\ b_1 \\ r \end{bmatrix} := \begin{bmatrix} \text{intercept(axis, R)} \\ \text{slope(axis, R)} \\ \text{corr(axis, R)} \end{bmatrix} \qquad \begin{bmatrix} b_0 \\ b_1 \\ r \end{bmatrix} := \begin{bmatrix} 0.3169 \\ 0.898 \\ 0.9949 \end{bmatrix}$$

FIGURE 3.9B
Quantile-quantile plot, ranges from a Weibull distribution.

Solutions

Exercise 3.1

Interpolation of the table of quantiles for the sample range yields

$$0.99 + \frac{4.358 - 4.120 \text{ (at } 0.99)}{4.424 \text{ (at } 0.995) - 4.120}(0.995 - 0.99) = 0.9939,$$

so the false alarm risk is 0.0061 as opposed to the expected 0.00135. The VBA function CDF_Range_Normal(4.358,3) returns 0.9942 for a false alarm risk of 0.0058.

Exercise 3.2

1. False alarm risk for

$$UCL = 1.964\sigma : \chi^2 = \frac{(n-1)UCL^2}{\sigma^2} = \frac{(5-1)1.964^2}{1^2} = 15.429.$$

$$= \text{CHIDIST}(15.429,4) \text{ returns an upper tail area of } 0.00389.$$

2. $$\chi^2 = \frac{(n-1)UCL^2}{\sigma^2} = \frac{(5-1)1.964^2}{1.275^2} = 9.49$$

which is the 0.95 quantile of the chi square distribution with 4 degrees of freedom. The power of this test to detect a 27.5% increase in the standard deviation is therefore 0.05, and the average run length is 20.

4

Nested Normal Distributions

Nested variation sources are often characteristic of batch processes. Levinson (1994) describes a process that coats semiconductor wafers with a photoresist whose thickness is the quality characteristic. Four or five measurements from a single wafer do not constitute a rational subgroup because they do not account for variation between wafers. The reference also cites a metallization chamber that processes more than a dozen silicon wafers at a time. A sample of four wafers from a batch is not a rational subgroup because it does not account for the variation that comes from the slightly different processing conditions for each batch. Figure 4.1 illustrates the different variation sources for an individual chip on a silicon wafer.

The distribution for each device is

$$X_{chip} \sim N\left(\mu_{process}, \sigma^2_{between\_batch} + \sigma^2_{within\_batch} + \sigma^2_{within\_wafer}\right)$$

and enormous confusion can result from failure to account for all three variation sources.

To begin with, individual pieces as opposed to their averages are in or out of specification. The denominator for the process capability or process performance index must therefore include all the variation sources. Incorrect calculation of control limits for statistical process control (SPC) charts will meanwhile result in inexplicable out-of-control signals on the x-bar chart while the chart for process variation does nothing unusual.

The issue of nested variation sources is, in fact, very similar to that of *gage repeatability and reproducibility* (R&R), in which measurement variation can come from the instrument (repeatability) or the way in which different inspectors read it (reproducibility). Isolation of the variance components is very straightforward when there are only two levels of nesting, for example, within batch and between batches. For two or more levels of nesting, StatGraphics offers Variance Components (under Compare/Analysis of Variance).

Variance Components: Two Levels of Nesting

If there are n observations per batch,

$$\sigma^2_{within\_batch} = MSE \quad \text{and} \quad \sigma^2_{between\_batch} = \frac{MST - MSE}{n}$$

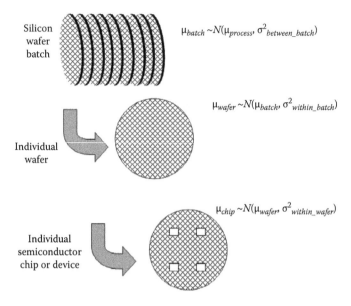

$\mu_{batch} \sim N(\mu_{process}, \sigma^2_{between\_batch})$

$\mu_{wafer} \sim N(\mu_{batch}, \sigma^2_{within\_batch})$

$\mu_{chip} \sim N(\mu_{wafer}, \sigma^2_{within\_wafer})$

FIGURE 4.1
Nested variation, semiconductor device manufacture.

where *MSE* is the mean square for errors and *MST* is the mean square for treatments from one-way analysis of variance (ANOVA). Each batch is therefore effectively a separate treatment and, if the F statistic for treatment effects is significant, the batch means are not identical.

The 50 batches of 4 that produced the unusual x-bar chart in the Introduction were simulated from a process with a mean of 100, a between-batch standard deviation of 4, and a within-batch standard deviation of 3 (Nested_normal tab in the simulated data worksheet). Table 4.1 shows the result from Excel's one-way analysis of variance procedure.

Then the estimated within-batch variance is 9.455, and the estimated between-batch variance is $\frac{54.466-9.455}{4} = 11.253$. StatGraphics' variance components

TABLE 4.1

One Way ANOVA Results, Batch Process

	Anova					
Source of Variation	**SS**	**df**	**MS**	**F**	**P-value**	**F crit**
Between Groups	2668.813	49	54.46558	5.760443	5.37E-17	1.439004
Within Groups	1418.266	150	9.455104			
Total	4087.079	199				

routine delivers the same results. Then

$$\sigma_{within\_batch} = 3.07$$

$$\sigma_{between\_batch} = 3.14$$

$$\sigma_{total} = \sqrt{9.455 + 11.253} = 4.55$$

The control limits for the average of n pieces from a single batch would then be

$$\mu \pm 3\sqrt{\frac{9.455}{n} + 11.253}$$

as opposed to

$$\mu \pm 3\sqrt{\frac{9.455}{n}},$$

which would be the estimate on the basis of the sample variation alone. This is why the x-bar chart for these applications often displays points outside of both control limits while the R or s chart behaves normally. The process performance index is meanwhile

$$\frac{USL - LSL}{6\sqrt{9.455 + 11.253}}$$

as opposed to

$$\frac{USL - LSL}{6\sqrt{9.455}}.$$

Batch processes are therefore highly undesirable from the perspective of both quality (due to the between-batch variance component) and statistical process control. Wortman (1991, XI-14) describes a Japanese tile manufacturing process that processed ceramic tiles in a kiln. The reason for unacceptable variation in the finished tiles' dimensions was obvious, as shown from the picture of the kiln. Tiles on the outside of each batch were exposed directly to radiative and convective heat transfer from the burners, while those inside probably had to rely on slower conductive heat transfer to reach the desired temperature. The variation in question was in fact not random but systematic, the kind that would require a multivariate control chart.

It is reasonable to expect other batch processes of this nature, such as batch food baking processes and heat treatment processes, to behave similarly.

Quality will depend on the product's sensitivity to the process variation. While the Japanese engineers conducted a designed experiment to make the tiles *robust* to temperature variation in the kiln, it is generally better to process units one at a time.

Batch versus Single Unit Processing

The previous section showed why batch processes complicate statistical process control while they reduce process capability. The tile manufacturing process would probably have worked a lot better with a belt furnace, in which each tile would have received almost identical amounts of heat. The barrier to implementation was the fact that the kiln cost $2 million in 1953 dollars, and a belt furnace with the necessary capacity would probably have been similarly expensive.

Batch processes are also the mortal enemy of lean manufacturing processes that seek to reduce inventory and cycle times. Heat treatment appears as "Herbie II," or the second capacity constraining operation, in Goldratt and Cox's *The Goal* (1992). Even if an assembly line produces one part at a time, the batch heat treatment process must wait to accumulate a full load for the furnace unless the furnace has excess capacity. This means that work accumulates as inventory and also waits, which adds to the cycle time. The parts in the finished heat treatment batch must then wait to go into the next single-unit process. Arnold and Faurote (1915, 86–87) show that Ford recognized the disadvantage of batch processes almost a century ago:

> This first [900 unit batch] annealing practice will soon be obsolete. A furnace now under construction is served by an endless chain moving up and down, which is fitted with pendulum blank-carriers to take the blanks individually as they come through the press die, carry them upward about 60 feet in the furnace uptake, giving ample heating time, and then carry the blanks downward 60 feet in the open air, giving plenty of cooling time before the blanks reach the oiling table.

Ford and Crowther (1930, 136) cite an observation by a blacksmith on the superiority of single-unit processing over batch-and-queue methods:

> The method in use [for heating of steel stock] when I was hired was to put twenty-five bars in the furnace at one time; then they were heated and forged and another lot put into the furnace, so there was always some time lost between "heats." But thought was applied to this business, and now the bars travel along rails through a regulated furnace which needs only one man to load, and the rest is done by an electrically operated automatic pusher. Thus there is a constant stream of correctly heated bars being supplied to the hammerman with a minimum of labour expended in the operation.

The continuous processes not only exposed the steel pieces to a relatively constant set of process conditions, they also delivered a constant stream of work as opposed to clumsy batches of inventory. The bottom-line conclusion is that continuous or single-unit processes are generally far more consistent with the objectives of lean and just-in-time manufacturing than corresponding batch processes. Their process capability is almost universally superior as well because single-unit processes do not have the built-in batch-to-batch variance component.

This chapter has shown how to compute control limits and process capability indices for batch processes. The next chapter will treat capability indices, including confidence intervals for the nonconforming fraction for nonnormal distributions.

Exercise

Exercise 4.1

A batch process has a within-batch variance of 4 square mils and a between-batch variance of 32 square mils. The process nominal is 500 mils, and the specification limits are [488,512] mils.

1. If the process variation is estimated from within-batch subgroup standard deviations or ranges (the incorrect assumption), what would the supplier report as the process capability index?
2. What is the real process capability index?
3. What are the control limits for an x-bar chart when samples of 4 are drawn from a single batch?
4. What are the control limits for the average of 8 measurements (2 per batch) from 4 batches?

Solution

Exercise 4.1

1. Incorrect assumption:

$$Cp = \frac{(512 - 488)}{6\sqrt{4}} = 2,$$

which corresponds to Six Sigma quality: 2 parts per billion nonconforming if the process is held at nominal.

2.

$$Cp = \frac{(512 - 488)}{6\sqrt{4 + 32}} = \frac{2}{3}$$

which corresponds to 4.55% nonconforming if the process is held at nominal.

3. The control limits for the average of four pieces from one batch are

$$500 \pm 3\sqrt{\frac{4}{4} + 32} = [482.76, 517.23].$$

4. The control limits for the average of eight pieces from four batches (two per batch) are

$$500 \pm 3\sqrt{\frac{4}{2} + \frac{32}{4}} = [490.51, 509.49].$$

5

Process Performance Indices

Process capability and process performance indices reflect the process' ability to perform within specifications, and customers often require periodic capability reports. It is therefore vital that organizations compute meaningful capabilities for nonnormal data.

Process Performance Index for Nonnormal Distributions

The Automotive Industry Action Group (2005, 142–143) sanctions two approaches and we recommend the first one shown below because there is no doubt as to the nonconforming fraction to which the index refers. These are technically process *performance* indices as opposed to process *capability* indices because the estimates of the distribution parameters come from maximum likelihood assessment of the entire dataset as opposed to subgroup statistics. As an example, a process performance index of 2 means one nonconformance per billion in the indicated tail of the distribution. The two approaches are

1. Compute the nonconforming fraction(s) p_L and p_U in each tail of the fitted nonnormal distribution. If there is only one specification limit, and this is frequently the case, compute only the relevant fraction.

   - Find the corresponding standard normal deviate (z statistic) for that nonconforming fraction: z such that $\Phi(z) = p$.
   - Divide the standard normal deviate's absolute value by 3 to get the process performance index:

$$P_p = \frac{|z|}{3}.$$

   As an example, suppose 0.00135 of a gamma distribution's upper tail is above the upper specification limit. The corresponding standard normal deviate is −3, and the process performance index is 1.

2. The other method is to divide the specification width by the estimated range of 99.73% of the data, that is, the 0.99865 quantile minus the 0.00135 quantile, to get the overall process performance index:

$$P_p = \frac{USL - LSL}{Q_{0.99865} - Q_{0.00135}}$$

- Note the similarity to

$$C_p = \frac{USL - LSL}{6\sigma},$$

which is the procedure's justification.

- An obvious problem is that it won't work for a process with a one-sided specification, which is frequent when the distribution is nonnormal. Another problem is that there is no guarantee that the resulting index will reflect the nonconforming fraction accurately.

The process capability or process performance index is effectively a statement or even a guarantee to the customer that the nonconforming fraction will not exceed a certain level. This chapter has shown so far how to compute a meaningful point estimate for the performance index from any distribution, but it leaves open the question as to this estimate's reliability. The Introduction showed that the confidence interval for the Pp of a normal distribution can be very wide indeed; the remainder of this chapter will therefore treat confidence limits for process capabilities.

Confidence Limits for Normal Process Performance Indices

Confidence limits for process performance indices correspond to *tolerance intervals*. The following quantities define a tolerance interval:

- The boundary (one-sided) or boundaries (two-sided) of the interval
- Confidence γ that fraction P of the population is within the interval

In the case of process capability or process performance, this translates into the supplier's statement to the customer that it is 100γ percent sure that the nonconforming fraction will not, in the absence of assignable cause variation, exceed $1 - P$.

The ability to state a confidence interval for the process performance indices should be a powerful sales tool. Suppose that a competitor's report claims Six Sigma process capability but also includes things like bimodal histograms (or no histograms at all) and/or a basis of 20 or 30 measurements (we have seen capability reports that cite even fewer measurements), while our report says:

> Past experience and/or the nature of this process suggests that the normal distribution is an appropriate model. 50 measurements were taken and, as shown by the attached histogram, chi square goodness-of-fit test, and quantile-quantile plot, the hypothesis that the distribution is normal cannot be rejected. The point estimate for the upper process performance index *PPU* is 2.0. We are 99% confident that this index is no less than 1.523, and that the process will therefore produce no more than 2.44 nonconformances per million pieces at the upper specification limit.

This makes it clear to the customer that (1) the process capability study's basis is an in-control process with a known distribution, and (2) the report accounts quantitatively and scientifically for the uncertainty in the process performance index. The customer will therefore have far more confidence in our company's report than that of the competitor, and this confidence should extend to our company's product as well.

Confidence Limits for P_p

The process performance index depends solely on the process's variance. When σ^2 is estimated from n data, the $100(1 - \alpha)\%$ confidence interval is

$$\frac{(n-1)}{\chi^2_{\frac{\alpha}{2};n-1}}s^2 \le \sigma^2 \le \frac{(n-1)}{\chi^2_{1-\frac{\alpha}{2};n-1}}s^2 \tag{5.1}$$

where the first subscript for χ^2 is the distribution's upper tail area (i.e., 1 minus the quantile). Then

$$\frac{\chi^2_{1-\frac{\alpha}{2};n-1}}{(n-1)}\frac{1}{s^2} \le \frac{1}{\sigma^2} \le \frac{\chi^2_{\frac{\alpha}{2};n-1}}{(n-1)}\frac{1}{s^2} \quad \Rightarrow \quad \sqrt{\frac{\chi^2_{1-\frac{\alpha}{2};n-1}}{(n-1)}}\frac{1}{s} \le \frac{1}{\sigma} \le \sqrt{\frac{\chi^2_{\frac{\alpha}{2};n-1}}{(n-1)}}\frac{1}{s}$$

and since $P_p = \dfrac{USL - LSL}{\sigma}$,

$$\hat{P}_p\sqrt{\frac{\chi^2_{1-\frac{\alpha}{2};n-1}}{(n-1)}} \le P_p \le \hat{P}_p\sqrt{\frac{\chi^2_{\frac{\alpha}{2};n-1}}{(n-1)}} \tag{5.2}$$

where the carat or "hat" (^) means the estimated quantity. In practice, only the lower confidence limit is likely to be of interest because the customer does not care how *few* nonconformances are present. It is easy to compute the confidence interval for P_p with tabulated values of chi square or Excel's CHIDIST function, but those for *PPU*, *PPL*, and P_{pk} involve uncertainty in both the process mean and standard deviation. The following methods can be checked against tabulated tolerance intervals for the normal distribution.

Confidence Limit for *PPL* and *PPU*

The confidence limit for *PPL* and *PPU* involves the same calculations that go into determination of *one-sided tolerance factors* for the normal distribution. Johnson and Kotz (1970, 204) show how to calculate the cumulative distribution for the noncentral *t* distribution where:

- $t' = k\sqrt{n}$ where *k* is the one-sided tolerance factor such that there is 100γ% confidence that fraction *P* of the population is either less than $\bar{x} + ks$ or more than $\bar{x} - ks$.
- The distribution's *noncentrality parameter* is $\delta = z_P\sqrt{n}$ where z_P is the standard normal deviate for *P*.
- The distribution's degrees of freedom are $\nu = n - 1$.

The cumulative density function for t' is then (Levinson, 1997a):

$$F(t') = \frac{1}{2^{\frac{\nu}{2}-1}\Gamma\left(\frac{\nu}{2}\right)} \int_0^\infty x^{\nu-1}\exp\left(-\frac{x^2}{2}\right)\Phi\left(\frac{t'x}{\sqrt{\nu}} - \delta\right)dx \qquad (5.3)$$

This is also the *confidence level* γ for the tolerance interval. Determination of the tolerance factor *k* begins with a given $\delta = z_P\sqrt{n}$, while determination of the confidence limit for the nonconforming fraction begins with *k* as a given.

Juran and Gryna (1988, AII.36) give 3.747 as the *k* factor for the one-sided 95% confidence interval for *P* = 99% of the population given a sample of 12 measurements. That is, we are 95% sure that 99% of the population is greater than the sample average minus 3.747 times its standard deviation, *or* less than the sample average plus 3.747 times its standard deviation. As shown in Figure 5.1, *F*(3.747) equals the confidence level of 0.95.

For larger sample sizes and degrees of freedom, it is necessary to move

$$\frac{1}{2^{\frac{\nu}{2}-1}\Gamma\left(\frac{\nu}{2}\right)}$$

inside the integral to avoid floating point errors. StatGraphics obtains the results in Table 5.1 where noncentrality $\delta = z_p\sqrt{n} = 2.326\sqrt{12} = 8.058$. Figure 5.2 illustrates the principle, and $t' = k\sqrt{n} = 3.747\sqrt{12} = 12.98$ on the axis.

StatGraphics can also return the tolerance factor for any desired confidence level and population fraction *P*. In this case, $\delta = z_P\sqrt{n}$ was provided

$\Phi(x) := cnorm(x)$

$$\begin{bmatrix} k \\ z_p \\ n \\ \infty \end{bmatrix} := \begin{bmatrix} 3.747 \\ 2.326 \\ 12 \\ 30 \end{bmatrix} \qquad \begin{bmatrix} v \\ \delta \\ t_{noncentral} \end{bmatrix} := \begin{bmatrix} n-1 \\ z_p \cdot \sqrt{n} \\ k \cdot \sqrt{n} \end{bmatrix} \qquad \begin{bmatrix} v \\ \delta \\ t_{noncentral} \\ \Phi(z_p) \end{bmatrix} = \begin{bmatrix} 11 \\ 8.058 \\ 12.98 \\ 0.99 \end{bmatrix}$$

$$F(t,\delta,v) := \frac{1}{2^{\frac{v}{2}-1} \cdot \Gamma\left(\frac{v}{2}\right)} \cdot \int_0^\infty x^{(v-1)} \cdot \exp\left(\frac{-x^2}{2}\right) \cdot \Phi\left(\frac{t \cdot x}{\sqrt{v}} - \delta\right) dx$$

$F(t_{noncentral}, \delta, v) = 0.95$ Confidence level for the tolerance interval

FIGURE 5.1
Cumulative noncentral *t* distribution.

for $P = 0.99$, and there are 11 degrees of freedom. The software's inverse cumulative distribution function (CDF) table returns 16.0479 as the 0.99 quantile of this noncentral *t* distribution. Dividing 16.0479 by the square root of 12 yields 4.633, the tabulated tolerance limit factor for $\gamma = 0.99$ and $P = 0.99$. In summary, the tolerance limit factor can be determined as follows:

1. Set $\delta = z_p \sqrt{n}$ where P is the population fraction.
2. Degrees of freedom $v = n - 1$.
3. Find the γ quantile (t') of the noncentral t distribution as defined by δ and v.
4. $k = \dfrac{t'}{\sqrt{n}}$

TABLE 5.1

StatGraphics CDF for Noncentral *t* Distribution

Distribution: Noncentral t		
Parameters:	**D. F.**	**Noncentrality**
Dist. 1	11	8.058
Dist. 2		
Cumulative Distribution		
Distribution: Noncentral *t*		
Lower Tail Area (<)		
Variable	**Dist. 1**	
12.98	0.950021	

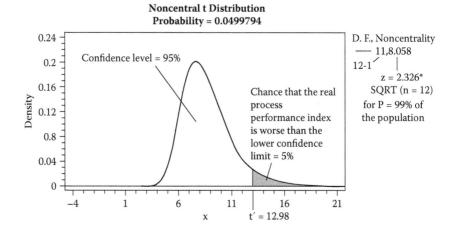

Noncentral t Distribution
Probability = 0.0499794

Confidence level = 95%

Chance that the real
process
performance index
is worse than the
lower confidence
limit = 5%

D. F., Noncentrality
—— 11,8.058
12-1

z = 2.326*
SQRT (n = 12)
for P = 99% of
the population

t′ = 12.98

FIGURE 5.2
Cumulative distribution, noncentral *t* distribution.

The problem statement for the process performance index will, however, begin with the desired confidence level (γ), the sample size (*n*), and the number of sample standard deviations (*k*) between the sample average and the specification limit. In other words, it starts with the tolerance factor as a given and requires computation of the corresponding population fraction. In this case,

1. $t' = k\sqrt{n}$
2. Degrees of freedom $v = n - 1$
3. Find δ to make the cumulative noncentral *t* distribution equal to the desired confidence level γ.
4. $z_p = \frac{\delta}{\sqrt{n}}$ and then divide by 3 to get the corresponding one-sided process performance index.

Suppose, for example, that there are 6.061 sample standard deviations (a little better than Six Sigma) between the average of 7 pieces and the upper specification limits. What is the lower 95% confidence level for the nonconforming fraction?

$$t' = k\sqrt{n} = 6.061\sqrt{7}$$

$$v = n - 1 = 7 - 1$$

and then find δ such that $F(t'|\delta) - 0.95 = 0$. Figure 5.3 shows a MathCAD bisection routine that does this.

$$\Phi(x) := \text{cnorm}(x)$$

$$\begin{bmatrix} \gamma \\ k \\ n \\ \infty \end{bmatrix} := \begin{bmatrix} .95 \\ 6.061 \\ 7 \\ 10 \end{bmatrix} \quad F(t, \delta, v) := \int_0^\infty \frac{x^{v-1}}{2^{\frac{v}{2}-1} \cdot \Gamma\left(\frac{v}{2}\right)} \cdot \exp\left(\frac{-x}{2}\right)^2 \cdot \Phi\left(\frac{t \cdot x}{\sqrt{v}} - \delta\right) dx$$

$$v := n-1 \quad f(\delta) := F(k \cdot \sqrt{n}, \delta, n-1) - \gamma \qquad \text{Function to be driven to zero}$$

$$\text{Bisect}(a, b) := \begin{vmatrix} x_L \leftarrow a \\ x_R \leftarrow b \\ \text{while } x_R - x_L > 0.0001 \\ \quad \begin{vmatrix} \text{test} \leftarrow f(x_L) \cdot f(x_R) \\ \text{TEMP} \leftarrow x_L \text{ if test} < 0 \\ x_R \leftarrow x_L \text{ if test} > 0 \\ x_L \leftarrow \text{TEMP if test} > 0 \\ x_L \leftarrow \frac{x_L + x_R}{2} \end{vmatrix} \\ \delta \leftarrow \frac{x_L + x_R}{2} \\ \text{Bisect} \leftarrow \Phi\left(\frac{\delta}{\sqrt{n}}\right) \end{vmatrix}$$

a and b are the limits for delta. a can be zero, and b must be high enough to ensure that the solution is in the interval.

Bisect (0, 10) = 0.999

FIGURE 5.3
95% Lower confidence limit for the nonconforming fraction.

Since $\delta = z_p \sqrt{n}$, the function returns $\Phi\left(\frac{\delta}{\sqrt{n}}\right) = 0.999$ as the confidence limit for the nonconforming fraction. This corresponds to $z_p = 3.09$, or $PPU = 1.03$. In other words, even though the point estimate for PPU is better than 2 ("Six Sigma"), the supplier is 95% sure only that PPU is 1.03 or higher.

Confidence Limit for *PPL* and *PPU* in Microsoft Excel

The Visual Basic for Applications (VBA) function Conf_PPL_Normal will perform the above calculations, and its arguments are k (sample standard deviations between the sample average and the lower or upper specification limit), sample size n, and the desired confidence level. It also works for *PPU*; the name indicates only that it is for PPL or PPU as opposed to P_{pk}. = Conf_PPL_Normal(6.061,7,0.95) returns 1.030 for the lower 95% confidence limit for *PPL* or *PPU*.

The question arises as to how well this program will work for larger sample sizes. Juran and Gryna (1988, AII.36–37) give 4.096 as the one-sided tolerance limit for 99.9% of the population and 99% confidence for a sample of

50. = Conf_PPL_Normal(4.096,50,0.99) returns a process performance index of 1.030, which corresponds to $P = 0.9990$.

Now suppose that a sample of 50 yields six standard deviations between the sample average and the specification limit of interest. $k = 6$ is well beyond the span of data in the table for $n = 50$, and it is important to know whether the function will deliver accurate results under these conditions. =Conf_PPL_Normal(6,50,0.99) returns 1.523 for the lower 99% confidence limit for *PPL* or *PPU*. This corresponds to 2.4358 defects per million opportunities, or $z_p = 4.5705$. (We are carrying what might look like an excessive number of significant figures here to allow comparison between results from three different algorithms, and rounding off would conceal slight differences in their results.)

- =CDF_Noncentral_t(6,50,0.999997564) returns 0.9900. This is not surprising, however, because the bisection routine used this function to get $P = 0.999997564$ in the first place.
- Integration in MathCAD also returns 0.9900.
- StatGraphics returns 0.9900 given $t' = 42.426$ (6 times the square root of 50) and $\delta = 32.318$ (4.5705 times the square root of 50).

The fact that the results from Conf_PPL_Normal agree with those from two other programs is encouraging, and VBA function Tolerance_One_Sided.bas generally reproduces the table of one-sided tolerance factors in Juran and Gryna (1988) to the nearest 0.003. The test results are in VBA.XLS. Notable exceptions for the 99% confidence limit are $k = 2.269$ versus the tabulated 2.296 for $n = 50$ and 95% of the population, and $k = 9.550$ versus 9.540 for $n = 6$ and $P = 0.999$. StatGraphics calculates $k = 2.269$ and 9.551, respectively, which suggests that the tabulated 2.296 might conceivably be a misprint with transposition of the last two digits.

It is not known how the function will perform for larger sample sizes, but Lieberman (1957) implies that the greatest discrepancies will occur for the smallest sample sizes: samples so small that they could never be acceptable as the basis for a meaningful capability study. Lieherman states that when the sample size exceeds 50, the tolerance factor may be approximated as follows:[1]

$$a = 1 - \frac{z_\gamma^2}{2(n-1)} \quad b = z_P^2 - \frac{z_\gamma^2}{n} \quad \text{and then} \quad k = \frac{z_P + \sqrt{z_P^2 - ab}}{a} \quad (5.4)$$

Example: 95% tolerance limit for 99% of the population. $z_{0.95} = 1.645$ and $z_{0.99} = 2.326$.

$$a = 1 - \frac{1.645^2}{2(50-1)} = 0.9724 \quad b = 2.326^2 - \frac{1.645^2}{50} = 5.356$$

$$k = \frac{2.326 + \sqrt{2.326^2 - 0.9724 \times 5.356}}{0.9724} = 2.854$$

The tabulated value is 2.863, and the difference is a little more than 0.3%. If k is given instead of P, this becomes an equation in one variable (z_p) and then the confidence limit for PPL or PPU is simply z_p divided by 3.

$$k = \frac{z_p + \sqrt{z_p^2(1-a) + a\frac{z_\gamma^2}{n}}}{a} \Rightarrow ak - z_p = \sqrt{z_p^2(1-a) + a\frac{z_\gamma^2}{n}}$$

$$a^2k^2 - 2akz_p + z_p^2 = z_p^2(1-a) + a\frac{z_\gamma^2}{n} \Rightarrow$$

$$az_p^2 - 2akz_p + \left(a^2k^2 - a\frac{z_\gamma^2}{n}\right) = 0$$

This is a simple quadratic equation for which no iterative solution is required:

$$z_p = \frac{ak - \sqrt{a^2k^2(1-a) + a^2\frac{z_\gamma^2}{n}}}{a} = k - \sqrt{k^2(1-a) + \frac{z_\gamma^2}{n}} \qquad (5.5)$$

As an example, suppose there are 2.854 sample standard deviations between the average of 50 measurements and the specification limit of interest. For the 95% confidence limit as shown above, $a = 0.9724$ and $z_\gamma = 1.645$ for the 95% confidence level.

$$z_p = 2.854 - \sqrt{2.854^2(1 - 0.9724) + \frac{1.645^2}{50}} = 2.327$$

which is approximately the standard normal deviate (2.326) for 99% of the population. Divide by 3 to get 0.776 as the lower 95% confidence limit for PPL or PPU. As n becomes even larger, the approximation should improve.

Although P_{pk} is the minimum of PPL and PPU, the lower confidence limit for P_{pk} is not the lesser confidence limit for PPL and PPU unless the process average differs substantially from the nominal. It is necessary to use the *bivariate noncentral* t *distribution* (Levinson, 1997a)[2] or the method described by Wald and Wolfowitz (1946).

Confidence Limit for P_{pk}

The procedure to find the lower confidence limit for P_{pk}, and also two-sided tolerance intervals for the normal distribution, is as follows: Let

$$Q(t', \delta, \nu, R) = \frac{\sqrt{2\pi}}{\Gamma\left(\frac{\nu}{2}\right)2^{\frac{\nu}{2}-1}} \int_R^\infty \Phi\left(\frac{t'x}{\sqrt{\nu}} - \delta\right)\left(\frac{1}{\sqrt{2\pi}}\exp\left(-\frac{x^2}{2}\right)\right)x^{\nu-1}dx \qquad (5.6)$$

where

$t' = k\sqrt{n}$ where k is again the one-sided tolerance factor such that fraction P of the population is either less than $\bar{x} + ks$ or more than $\bar{x} - ks$. In this application, there will however be a t' for each specification limit.

The distribution's *noncentrality parameter* is $\delta = z_p\sqrt{n}$ where z_p is the standard normal deviate for P. There will also be a δ for each specification limit.

$$R = \frac{\delta_1 + \delta_2}{t'_1 + t'_2}\sqrt{n} = \frac{\delta_1 + \delta_2}{k_1 + k_2}$$

or just $\frac{\delta}{k}$ when $\delta_1 = \delta_2$ and $t'_1 = t'_2$

Degrees of freedom $v = n - 1$

To find the lower confidence limit for P_{pk}, solve the following equation for $\delta_1 = \delta_2$ to avoid an infinite number of possible solutions. To find the two-sided tolerance interval for a given population fraction, let $t'_1 = t'_2$ and find $\delta_1 = \delta_2 = z_p\sqrt{n}$ such that:

$$Q(t'_1, \delta_1, v, R) + Q(t'_2, \delta_2, v, R) - 1 = \gamma$$

for which Papp (1992) provides the following equation.

$$Q(t'_1, t'_2, \delta_1, \delta_2, v, R) = \frac{\sqrt{2\pi}}{\Gamma(\frac{v}{2})2^{\frac{v}{2}-1}}\int_R^\infty \left[\Phi\left(\frac{t'_1 x}{\sqrt{v}} - \delta_1\right) + \Phi\left(\frac{t'_2 x}{\sqrt{v}} - \delta_2\right) - 1\right]$$

$$\times \left(\frac{1}{\sqrt{2\pi}}\exp\left(-\frac{x^2}{2}\right)\right)x^{v-1}dx$$

(5.6a)

$t'_1 = t'_2$ and $\delta_1 = \delta_2$ simplifies this to:

$$Q(t', \delta, v, R) = \frac{1}{\Gamma(\frac{v}{2})2^{\frac{v}{2}-1}}\int_R^\infty \left(2\Phi\left(\frac{t'x}{\sqrt{v}} - \delta\right) - 1\right)\times\exp\left(-\frac{x^2}{2}\right)x^{v-1}dx \quad (5.6b)$$

Chou, Owen, and Borrego (1990) obtain 0.723 as the 95% lower confidence limit for P_{pk} given a sample of 30 and a point estimate of $P_{pk} = 1$. The algorithm in Figure 5.4 reproduces this result exactly.

Juran and Gryna (1988, AII.37) give 3.350 as the two-sided tolerance factor for 99% of the population with 95% confidence when the sample size is 30, but the routine shown above returns 99.28% of the population as the lower confidence limit. Furthermore, the solution for k such that $F(k) = 0.95$ where

$$\begin{bmatrix} n \\ \gamma \\ k \end{bmatrix} := \begin{bmatrix} 30 \\ .95 \\ 3 \end{bmatrix} \qquad \begin{bmatrix} \upsilon \\ t \end{bmatrix} := \begin{bmatrix} n-1 \\ k\sqrt{n} \end{bmatrix} \qquad \Phi(x) := \text{cnorm}(x)$$

$$F(\delta) := \int_{\frac{\delta}{k}}^{15} \frac{1}{2^{\frac{\upsilon}{2}-1}\cdot\Gamma\left(\frac{\upsilon}{2}\right)}\left(2\cdot\Phi\left(\frac{t\cdot x}{\sqrt{\upsilon}}-\delta\right)-1\right)\cdot\exp\left(\frac{-x^2}{2}\right)\cdot x^{\upsilon-1}dx$$

Cumulative distribution when the sample average is halfway between the specification limits

$f(\delta) := F(\delta) - \gamma$ Function to be driven to zero

$$\text{Bisect(b)} := \begin{vmatrix} x_L \leftarrow 0 \\ x_R \leftarrow b \\ \text{while } x_R - x_L > 0.0001 \\ \quad \begin{vmatrix} \text{test} \leftarrow f\langle x_L\rangle \cdot f\langle x_R\rangle \\ \text{TEMP} \leftarrow x_L \text{ if test} < 0 \\ x_R \leftarrow x_L \text{ if test} > 0 \\ x_L \leftarrow \text{TEMP if test} > 0 \\ x_L \leftarrow \dfrac{x_L + x_R}{2} \end{vmatrix} \\ \delta \leftarrow \dfrac{x_L + x_R}{2} \\ \text{Bisect} \leftarrow \dfrac{\delta}{\sqrt{n}} \end{vmatrix}$$

$$\frac{\text{Bisect}(25)}{3} = 0.723$$

FIGURE 5.4
95% lower confidence limit for P_{pk}.

$\delta = z_{0.99}\sqrt{30}$ yields 3.20 as the tolerance factor. The reason for this discrepancy appears to be the fact that the table in Juran and Gryna is based on the following procedure (Wald and Wolfowitz, 1946):

$$k = \sqrt{\frac{\upsilon}{\chi^2_{\upsilon;\gamma}}}r \text{ where } P = \Phi\left(\frac{1}{\sqrt{n}}+r\right) - \Phi\left(\frac{1}{\sqrt{n}}-r\right) \text{ and } \upsilon = n-1 \qquad (5.7)$$

As shown in Figure 5.5, this algorithm reproduces the 3.350 tolerance factor; it also reproduces, to within one thousandth, the column of 95% tolerance factors for 99% of the population as shown in Juran and Gryna (1988).

VBA function Conf_PPK_Wald returns the lower confidence limit for P_{pk} given k, the sample size, and the desired confidence level. =Conf_PPK_Wald(3.35,30,0.95) returns 0.7755 for the 95% lower confidence limit for P_{pk}, which corresponds to $P = 0.9900$. In other words, it is consistent with the table of two-sided tolerance factors in Juran and Gryna (1988). VBA function

2 - Sided Tolerance Factors, Wald and Wolfowitz

$$\Phi(x) := \text{cnorm}(x) \qquad f(r,n,P) := \Phi\left(\frac{1}{\sqrt{n}} + r\right) - \Phi\left(\frac{1}{\sqrt{n}} - r\right) - P$$

$$k(n, P, \gamma) := \begin{array}{|l} x_L \leftarrow 0 \\ x_R \leftarrow 6 \\ \text{while } x_R - x_L > 0.0001 \\ \quad \begin{array}{|l} \text{test} \leftarrow f\langle x_L, n, p\rangle \cdot f\langle x_R, n, p\rangle \\ \text{TEMP} \leftarrow x_L \text{ if test} < 0 \\ x_R \leftarrow x_L \text{ if test} > 0 \\ x_L \leftarrow \text{TEMP if test} > 0 \\ x_L \leftarrow \dfrac{x_L + x_R}{2} \end{array} \\ r \leftarrow \dfrac{x_L + x_R}{2} \\ k \leftarrow r \cdot \sqrt{\dfrac{n-1}{\text{qchisq}(1-\gamma, n-1)}} \end{array} \qquad k(30, .99, .95) = 3.35$$

FIGURE 5.5
Wald and Wolfowitz procedure for 2-sided tolerance factor.

Tolerance_Wald reproduces the entire table of two-sided tolerance factors from this reference to the nearest 0.001 (nearest 0.002 when the tabulated value exceeds 100), and the test results are in the VBA.XLS spreadsheet.

Kushler and Hurley (1992) meanwhile offer the following approximation for the lower confidence bound for C_{pk} (technically P_{pk} if the estimate of the standard deviation is that of the entire dataset as opposed to the subgroup statistics).[3]

$$\hat{C}_{pk}\left[1 - z_\gamma \left[\frac{n-1}{n-3} - \frac{n-1}{2}\left(\frac{\Gamma\left(\frac{n-2}{2}\right)}{\Gamma\left(\frac{n-1}{2}\right)}\right)^2\right]^{\frac{1}{2}}\right] \tag{5.8}$$

The same reference also cites a lower confidence bound of

$$\hat{C}_{pk} - z_\gamma\sqrt{\frac{1}{9n} + \frac{\hat{C}_{pk}^2}{2n-2}} \tag{5.9}$$

There are, in summary, four ways to solve for the lower confidence limit for P_{pk}.

(1) Bivariate Noncentral t Distribution Method

Given k sample standard deviations between the process average and the specification limits, find δ such that

$$\frac{1}{\Gamma\left(\frac{v}{2}\right)2^{\frac{v}{2}-1}} \int_R^\infty \left(2\Phi\left(\frac{t'x}{\sqrt{v}}-\delta\right)-1\right) \times \exp\left(-\frac{x^2}{2}\right) x^{v-1} dx - \gamma = 0$$

(5.10)

where $\delta = z_p\sqrt{n}$ $\quad t' = k\sqrt{n}$ and $R = \dfrac{\delta}{k}$

It is then straightforward to convert δ into the upper confidence limit for the nonconforming fraction and the lower confidence limit for P_{pk}. This method reproduces the result in the example from Chou, Owen, and Borrego (1990).

(2) Wald and Wolfowitz Procedure

Given k sample standard deviations between the process average and the specification limits:

$$r = \sqrt{\frac{\chi^2_{v;\gamma}}{v}}k \quad \text{and then find} \quad P = \Phi\left(\frac{1}{\sqrt{n}}+r\right) - \Phi\left(\frac{1}{\sqrt{n}}-r\right)$$

(5.11)

This procedure is considerably older (and simpler) than the first, but the fact that it reproduces the table from Juran and Gryna (1988, AII.37) and Beyer (1991, 116–117) is encouraging.

(3) An Approximation That Improves with Increasing Sample Size

$$\hat{P}_{pk}\left[1-z_\gamma\left[\frac{n-1}{n-3}-\frac{n-1}{2}\left(\frac{\Gamma\left(\frac{n-2}{2}\right)}{\Gamma\left(\frac{n-1}{2}\right)}\right)^2\right]^{\frac{1}{2}}\right]$$

(5.12)

(4) Another Approximation That Improves with Larger Samples

$$\hat{P}_{pk} - z_\gamma\sqrt{\frac{1}{9n}+\frac{\hat{C}^2_{pk}}{2n-2}}$$

(5.13)

StatGraphics Centurion uses this Equation (5.13) to estimate the lower confidence limit for C_{pk} and P_{pk}.

Example:

Given 100 measurements with an average of 59.962, standard deviation 2.124, and specification limits [49.342,70.582], what is the lower 95% confidence limit for P_{pk}?

1. Use of the bivariate noncentral t distribution yields $z_P = 4.321$, which corresponds to $P_{pk} = 1.440$.
2. Use of the Wald-Wolfowitz procedure, =Conf_Ppk_Wald(5,100,0.95), returns 1.412.

3. $\hat{P}_{pk}\left[1 - z_\gamma\left[\dfrac{n-1}{n-3} - \dfrac{n-1}{2}\left(\dfrac{\Gamma\left(\frac{n-2}{2}\right)}{\Gamma\left(\frac{n-1}{2}\right)}\right)^2\right]^{\frac{1}{2}}\right]$ yields 1.468.

4. $\hat{P}_{pk} - z_\gamma\sqrt{\dfrac{1}{9n} + \dfrac{\hat{C}_{pk}^2}{2n-2}}$ yields 1.464, as does StatGraphics.

This section has shown how to determine lower confidence limits for the normal distribution's process performance indices, and the VBA functions facilitate the calculations in Microsoft Excel. This enables the practitioner to quote not only process performance indices as is often required by customers, but also confidence limits on the indices and corresponding nonconforming fractions. The next section will address the same issue for nonnormal distributions.

Confidence Limits for Nonnormal Process Performance Indices

The issue of confidence limits for nonnormal capability indices ties in very closely with confidence limits for the survivor function in reliability testing. In the latter case, it is desirable to identify not only a point estimate of the population that will survive to a given time (or number of cycles, or other units of use) but also a confidence limit for the surviving fraction. If "specification limit" replaces "given time" in this discussion, the same methods can determine an upper confidence limit on the nonconforming fraction that corresponds to a lower confidence limit for the process performance index.

Lawless (1982, 211) states that it is extremely difficult to determine confidence limits for the survivor function of the gamma distribution, although the same reference has some recommendations begining on page 216. It also describes (pp. 174–175) a general procedure for the Weibull distribution that relies on the large-sample behavior of the *likelihood ratio statistic*. The importance of large-sample behavior cannot be overemphasized, but remember

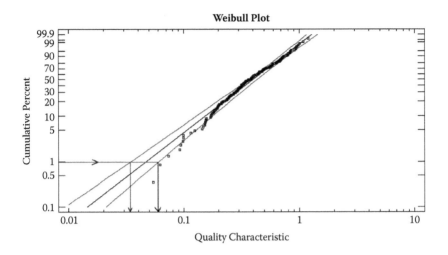

FIGURE 5.6A
95% confidence interval for 1% of the population.

that process performance indices whose basis consists of only ten or twenty measurements are not particularly meaningful.

The procedure's first application is determination of the confidence interval for the time to which a given fraction of the population will survive, and the second is the confidence interval for the fraction of the population that will survive to a given time. The latter approach can also find the confidence interval for the fraction of the population that will be inside a specification limit, and therefore a process performance index. Figure 5.6 (StatGraphics Centurion, Weibull plot with 95% confidence intervals) illustrates these applications for the Weibull distribution whose parameters were found to be $\beta = 1.9686$ and $\theta = 0.48214$, and whose specification limits are [0.03,1.30]. (The limits for this data set were [0.02,1.30] in Chapter 2 but it will be easier to illustrate the idea for a lower specification of 0.03. The data are in Weibull_b2_theta_0.5 in the simulated data spreadsheet.)

In Figure 5.6A, the first percentile of the population was selected because this portion of the figure is easier to see than, for example, the 90th or 99th percentile. The figure shows that there is 95% confidence that the first percentile is between about 0.033 and 0.060. Since this is a two-sided confidence limit, there is 97.5% confidence that the first percent of the population will survive past 0.033 cycles, hours, or whatever the life measurement happens to be. There is 97.5% confidence that the first percent of the population will not survive past 0.060.

Figure 5.6B begins with the premise that the lower specification limit is 0.03. The figure shows 97.5% confidence that no more than about 0.8% of the population will be below the lower specification limit, and this would

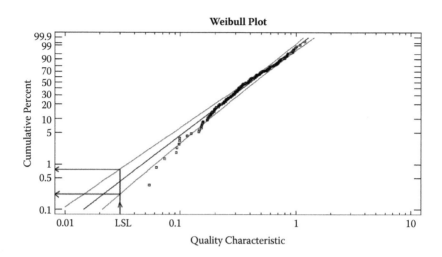

FIGURE 5.6B
95% confidence interval for nonconforming fraction.

become the basis of a confidence limit for *PPL*. The next sections describe how to calculate these confidence intervals.

Confidence Interval for Population Fraction *P*, Weibull Distribution

Let t_p be the point estimate of the time to which fraction p of the population survives. Given *p1* fitted parameters—two for the gamma and Weibull distributions—the likelihood ratio statistic behaves as follows for a test of the hypothesis $t_p = Q_0$ (Lawless, 1982, 174–175). The parameters of the Weibull distribution illustrate the procedure.

$$\Lambda = -2 \ln L(\tilde{\beta}, \tilde{\theta}) + 2 \ln L(\hat{\beta}, \hat{\theta}) \sim \chi^2_{v=p1-1}$$

where $\tilde{\beta}, \tilde{\theta}$ are calculated on the basis of Q_0

and $\hat{\beta}, \hat{\theta}$ are the maximum likelihood estimates

In other words, Q_0 as the γ confidence limit for the survival time of fraction p of the population must meet the following conditions:

(1) $\Lambda = -2 \ln L(\tilde{\beta}, \tilde{\theta}) + 2 \ln L(\hat{\beta}, \hat{\theta}) = \gamma$ quantile of χ^2_1

(2) $\tilde{\beta}$ maximizes $\ln L(\tilde{\beta}, \tilde{\theta}(Q_0, \tilde{\beta}))$ (Set 5.14)

(3) $\tilde{\theta} = \dfrac{Q_0}{(-\ln(1-p))^{1/\tilde{\beta}}}$

The second condition is significant because

$$\frac{\partial}{\partial \tilde{\beta}} \ln L(\tilde{\beta}, \tilde{\theta}(Q_0, \tilde{\beta})) = 0$$

as opposed to $\dfrac{\partial}{\partial \tilde{\beta}} \ln L(\tilde{\beta}, \tilde{\theta}) = 0$ and $\dfrac{\partial}{\partial \tilde{\theta}} \ln L(\tilde{\beta}, \tilde{\theta}) = 0$

The third condition is redundant because it is built into the solution procedure. The real problem consists of a set of two equations and two variables (Q_0 and $\tilde{\beta}$), with $\tilde{\theta}$ dependent on the latter quantities.

The γ confidence interval for t_P is such that, given calculation of $\tilde{\beta}, \tilde{\theta}$ from Q_0, $\Lambda \leq \chi^2_{\gamma,1}$ (the γ quantile of chi square with one degree of freedom). This requires a two-stage iterative solution.

1. Select a value for Q_0, the candidate value for the confidence limit. Bisection of an interval $[a,b]$ is certain to obtain an answer if the answer lies in this interval.
2. Calculate $\tilde{\beta}$ and $\tilde{\theta}$ as shown below.
3. Compute the likelihood ratio statistic $\Lambda = -2 \ln L(\tilde{\beta}, \tilde{\theta}) + 2 \ln L(\hat{\beta}, \hat{\theta})$.
4. Iterate Q_0 (e.g., with bisection) until convergence occurs.

Recall that quantile q of the Weibull distribution is $F^{-1}(q) = \theta(-\ln(1-q))^{1/\beta} + \delta$. If the threshold parameter is ignored (it can be deducted from the data and Q_0),

$$F^{-1}(p) = Q_0 = \theta(-\ln(1-p))^{1/\beta} \Rightarrow$$

$$\theta = \frac{Q_0}{(-\ln(1-p))^{1/\beta}}$$

Substitute this into the maximum likelihood function, and then set the partial derivative for β equal to zero.

$$\ln L(X) = \sum_{i=1}^{n} \ln \beta - \beta \ln \theta + (\beta - 1) \ln(x_i) - \left(\frac{x_i}{\theta}\right)^{\beta}$$

$$= \sum_{i=1}^{n} \ln \beta - \beta \ln Q_0 + \beta \frac{1}{\beta} \ln(-\ln(1-p)) + (\beta-1)\ln(x_i) + \ln(1-p)\left(\frac{x_i}{Q_0}\right)^{\beta}$$

$$\frac{\partial \ln L(X)}{\partial \beta} = \frac{n}{\beta} - n \ln Q_0 + \sum_{i=1}^{n} \ln x_i + \ln(1-p)\sum_{i=1}^{n}\left(\frac{x_i}{Q_0}\right)^{\beta} \ln\left(\frac{x_i}{Q_0}\right) = 0$$

Solve to obtain $\tilde{\beta}$ and then $\tilde{\theta} = \dfrac{Q_0}{(-\ln(1-p))^{1/\tilde{\beta}}}$

$$f(\beta) := \frac{\sum_{i=1}^{n}(x_i)^\beta \cdot \ln(x_i)}{\sum_{i=1}^{n}(x_i)^\beta} - \frac{1}{\beta} - \frac{1}{n}\sum_{i=1}^{n}\ln(x_i)$$ Function to be driven to zero

Range to bisect $\begin{bmatrix} a \\ b \end{bmatrix} := \begin{bmatrix} 1 \\ 10 \end{bmatrix}$ $f(a)\cdot f(b) = -0.662$
Test to make sure a
solution is in the range
to be bisected

Graph of the function $j := 1..101$ $y_j := a + \frac{(j-1)}{100}\cdot(b-a)$

$$\text{Bisect}(a, b) := \left| \begin{array}{l} x_L \leftarrow 0 \\ x_R \leftarrow b \\ \text{while } x_R - x_L > 0.000001 \\ \quad \left| \begin{array}{l} \text{test} \leftarrow f\langle x_L \rangle \cdot f\langle x_R \rangle \\ \text{TEMP} \leftarrow x_L \ \text{if test} < 0 \\ x_R \leftarrow x_L \ \text{if test} > 0 \\ x_L \leftarrow \text{TEMP if test} > 0 \\ x_L \leftarrow \dfrac{x_L + x_R}{2} \end{array} \right. \\ \text{Bisect} \leftarrow \dfrac{x_L + x_R}{2} \end{array} \right.$$

$\beta := \text{Bisect}(a, b)$

$$\theta := \left[\frac{1}{n}\cdot\sum_{i=1}^{n}(x_i)^\beta \right]^{\frac{1}{\beta}}$$

$$\begin{bmatrix} \beta \\ \theta \end{bmatrix} = \begin{bmatrix} 1.9686 \\ 0.4821 \end{bmatrix}$$

$$\text{LogL} := n\cdot\ln(\beta) - n\cdot\beta\cdot\ln(\theta) + (\beta-1)\cdot\sum_{i=1}^{n}\ln(x_i) - \sum_{i=1}^{n}\left(\frac{x_i}{\theta}\right)^\beta \quad \text{LogL} = 26.907$$

FIGURE 5.7A
Confidence interval for first percentile, Weibull distribution.

The next step is to compute $\ln L(\tilde{\beta}, \tilde{\theta})$, the log likelihood function whose basis is $\tilde{\beta}, \tilde{\theta}$, and then to compute the likelihood ratio statistic Λ. Iterate Q_0 until $\Lambda = \chi^2_{1,\gamma}$ where γ is the desired two-sided confidence level. As shown in Figure 5.7, there will be two solutions, and therefore two intervals that will require separate bisections if both ends of the confidence interval are required. The 95% confidence interval for the first percentile, [0.0344, 0.0606], is consistent with StatGraphics' graphical output, and the program's documentation shows that it does in fact use the method shown previously.

The procedure begins by fitting the Weibull distribution to the dataset, and it also calculates the log likelihood function for the best-fit parameters. LogL is $\ln L(\hat{\beta}, \hat{\theta})$ for subsequent calculation of the likelihood ratio statistic.

Next is a function to compute the likelihood ratio for Q_0.

A graph of the likelihood ratio versus Q_0 shows two solution points such that the likelihood ratio equals $\chi^2 = 3.841$. These are the lower and upper confidence limits, as shown in Figure 5.7.

$$fl\left(\beta, Q_0, p\right) := \frac{n}{\beta} - n \cdot \ln\left(Q_0\right) + \sum_{i=1}^{n} \ln(x_i) + \ln(1-p) \cdot \sum_{i=1}^{n} \left(\frac{x_i}{Q_0}\right)^{\beta} \cdot \ln\left(\frac{x_i}{Q_0}\right)$$

Likelihood_Ratio $(Q_0, p) :=$ | $x_L \leftarrow a$
| $x_R \leftarrow b$
| while $x_R - x_L > 0.000001$
| | test $\leftarrow fl\langle x_L, Q_0, p\rangle \cdot fl \langle x_R, Q_0, p\rangle$
| | TEMP $\leftarrow x_L$ if test < 0
| | $x_R \leftarrow x_L$ if test > 0
| | $x_L \leftarrow$ TEMP if test > 0
| | $x_L \leftarrow \dfrac{x_L + x_R}{2}$
| $\beta \leftarrow \dfrac{x_L + x_R}{2}$
| $\theta \leftarrow \dfrac{Q_0}{(-\ln(1-p))^{1/\beta}}$
| logL_tilde $\leftarrow (n \cdot \ln(\beta) - n \cdot \beta \cdot \ln(\theta)) \ldots$
| $\qquad + (\beta - 1) \cdot \sum_{i=1}^{n} \ln(x_i) - \sum_{i=1}^{n} \left(\dfrac{x_i}{\theta}\right)^{\beta}$
| Likelihood_Ratio $\leftarrow -2 \cdot$ logL_tilde $+ 2 \cdot$ LogL

FIGURE 5.7B
Confidence interval for first percentile, Weibull distribution.

Bisection requires that the bisected interval contain only one solution, and this graph shows that there are two. Bisection of the interval [0.03,0.05] yields 0.0344 for the lower confidence limit and, as shown in Figure 5.7D, bisection of [0.05,0.07] yields 0.0606 for the upper confidence limit.

The same algorithm reproduces the results for the half-life ($t_{0.50}$) for the placebo data from Lawless (1982, 175–176). The MathCAD program gets [4.755, 10.256] versus [4.75, 10.3] from the reference.

Confidence Interval for Survivor Function (or Nonconforming Fraction)

In this case, the objective is to get a γ confidence interval (or one-sided limit) for the nonconforming fraction p at specification limit Q. This corresponds to the confidence limit for $F(Q)$ at the lower specification limit and $1 - F(Q)$, or the reliability survivor function $S(Q)$, at the upper specification limit. In the latter case, find S_0 such that, given calculation of $\hat{\beta}, \hat{\theta}$ from S_0, $\Lambda = \chi^2_{\gamma,1}$. S_0 as the γ confidence limit for the survival function at time (or specification limit

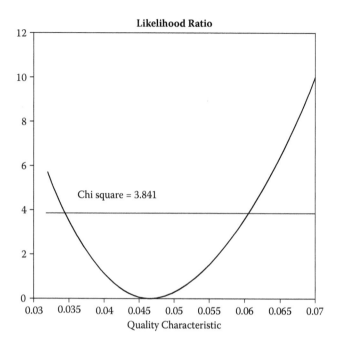

FIGURE 5.7C
Confidence interval for first percentile, Weibull distribution.

Bisection for Q

Function to be driven to zero:

$f\_Q(Q,p) := \text{Likelihood\_Ratio}(Q,p) - \text{ChiSquare}$

$$\text{Bisect\_Q}(Q_L, Q_R, p) := \begin{array}{|l} x_L \leftarrow ab \\ x_R \leftarrow aa \\ \text{while } x_R - x_L > 0.000001 \\ \quad \begin{array}{|l} \text{test} \leftarrow f\_Q(x_L, p) \cdot f\_Q\langle x_R, p\rangle \\ \text{TEMP} \leftarrow x_L \text{ if test} < 0 \\ x_R \leftarrow x_L \text{ if test} > 0 \\ x_L \leftarrow \text{TEMP if test} > 0 \\ x_L \leftarrow \dfrac{x_L + x_R}{2} \end{array} \\ \text{Bisect} \leftarrow \dfrac{x_L + x_R}{2} \end{array}$$

$\text{Bisect\_Q}(.05, .07, .01) = 0.0606$

FIGURE 5.7D
Confidence interval for first percentile, Weibull distribution.

Q) must meet the following conditions:

(1) $\Lambda = -2\ln L(\tilde{\beta}, \tilde{\theta}) + 2\ln L(\hat{\beta}, \hat{\theta}) = \gamma$ quantile of χ_1^2

(2) $\tilde{\beta}$ maximizes $\ln L(\tilde{\beta}, \tilde{\theta}(S_0, \tilde{\beta}))$ (Set 5.15)

(3) $\tilde{\theta} = \dfrac{Q}{(-\ln S_0)^{1/\tilde{\beta}}}$

The third condition is again redundant, and the actual system consists of two equations and two unknowns (S_0 and $\tilde{\beta}$). This equation system also requires a two-stage iterative solution.

1. Select a value for S_0, the candidate value for the confidence limit. Bisection of an interval $[a,b]$ is certain to obtain an answer if the answer lies in this interval.
2. Calculate $\tilde{\beta}$ and $\tilde{\theta}$ as shown below.
3. Compute the likelihood ratio statistic.
4. Iterate S_0 (e.g., with bisection) until convergence occurs.

For the two-parameter Weibull distribution,

$$F(Q) = 1 - \exp\left(-\left(\frac{Q}{\theta}\right)^{\beta}\right) \Rightarrow S(Q) = \exp\left(-\left(\frac{Q}{\theta}\right)^{\beta}\right)$$

and then θ

$$= \frac{Q}{(-\ln S_0)^{1/\beta}}$$

Substitute this into the maximum likelihood function, and then set the partial derivative for β equal to zero.

$$\ln L(X) = \sum_{i=1}^{n} \ln \beta - \beta \ln \theta + (\beta - 1)\ln(x_i) - \left(\frac{x_i}{\theta}\right)^{\beta}$$

$$= \sum_{i=1}^{n} \ln \beta - \beta \ln Q + \beta \frac{1}{\beta}\ln S_0 + (\beta - 1)\ln(x_i) + \ln S_0 \left(\frac{x_i}{Q}\right)^{\beta}$$

$$\frac{\partial \ln L(X)}{\partial \beta} = \frac{n}{\beta} - n\ln Q + \sum_{i=1}^{n} \ln x_i + \ln S_0 \sum_{i=1}^{n} \left(\frac{x_i}{Q}\right)^{\beta} \ln\left(\frac{x_i}{Q}\right) = 0$$

Solve to obtain $\tilde{\beta}$ and then $\tilde{\theta} = \dfrac{Q}{(-\ln S_0)^{1/\tilde{\beta}}}$

Drive the following function to 0 to get beta_tilde

$$fl(\beta, S_0, Q) := \frac{n}{\beta} - n \cdot \ln(Q) + \sum_{i=1}^{n} \ln(x_i) + \ln(S_0) \cdot \sum_{i=1}^{n} \left(\frac{x_i}{Q}\right)^{\beta} \cdot \ln\left(\frac{x_i}{Q}\right)$$

Likelihood_Ratio $(S_0, Q) :=$
$\begin{array}{l}
x_L \leftarrow a \\
x_R \leftarrow b \\
\text{while } x_R - x_L > 0.000001 \\
\quad \left| \begin{array}{l}
\text{test} \leftarrow fl(x_L, S_0, Q) \cdot fl(x_R, S_0, Q) \\
\text{TEMP} \leftarrow x_L \text{ if test} < 0 \\
x_R \leftarrow x_L \text{ if test} > 0 \\
x_L \leftarrow \text{TEMP if test} > 0 \\
x_L \leftarrow \dfrac{x_L + x_R}{2}
\end{array} \right. \\
\beta \leftarrow \dfrac{x_L + x_R}{2} \\
\theta \leftarrow \dfrac{Q}{(-\ln(S_0))^{\frac{1}{\beta}}} \\
\text{logL_tilde} \leftarrow (n \cdot \ln(\beta) - n \cdot \beta \cdot \ln(\theta))\ldots \\
\quad + (\beta - 1) \cdot \sum_{i=1}^{n} \ln(x_i) - \sum_{i=1}^{n} \left(\dfrac{x_i}{\theta}\right)^{\beta} \\
\text{Likelihood_Ratio} \leftarrow -2 \cdot \text{logL_tilde} + 2 \cdot \text{LogL}
\end{array}$

FIGURE 5.8A
Confidence interval for survivor function, Weibull distribution.

Compute the likelihood function whose basis is $\tilde{\beta}, \tilde{\theta}$, and then find the likelihood ratio statistic. Iterate S_0 until $\Lambda = \chi^2_{1,\gamma}$ where γ is the desired one-sided confidence level.

The following MathCAD algorithm is similar to the previous one; in fact, the first part, which finds $\ln L(\hat{\beta}, \hat{\theta})$ for subsequent calculation of the likelihood ratio statistic, is identical. Figure 5.8 shows the subsequent calculations for the Weibull distribution whose fitted parameters are $\beta = 1.9686$ and $\theta = 0.48214$, and whose lower specification limit is assumed to be 0.03 for this exercise. Note that Q (the specification limit or equivalent) is given in this case, and confidence limits for the survivor function (or nonconforming fraction) are to be determined.

A plot of the likelihood ratio versus the survivor function (fraction of the population above LSL = 0.03) shows that there will again be two solutions. Only the lower number is of practical interest because it corresponds to the lower confidence limit for *PPL*.

Bisection of the interval [0.99,0.994] returns a lower confidence limit of 0.9921 for the survivor function or, in this case, the fraction of the population

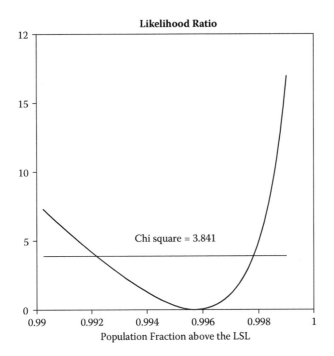

FIGURE 5.8B
Confidence interval for survivor function, Weibull distribution.

Bisection for S

Function to be driven to zero:

$f\_S\ (S_0, Q) := Likelihood\_Ratio\ (S_0, Q)\ -\ ChiSquare$

$$Bisect\_S(S_L, S_R, Q) := \begin{vmatrix} x_L \leftarrow S_L \\ x_R \leftarrow S_R \\ \text{while } x_R - x_L > 0.000001 \\ \quad \begin{vmatrix} test \leftarrow f\_S(x_L, Q) \cdot f\_S\ (x_R, Q) \\ TEMP \leftarrow x_L \text{ if test} < 0 \\ x_R \leftarrow x_L \text{ if test} > 0 \\ x_L \leftarrow TEMP \text{ if test} > 0 \\ x_L \leftarrow \dfrac{x_L + x_R}{2} \end{vmatrix} \\ Bisect \leftarrow \dfrac{x_L + x_R}{2} \end{vmatrix}$$

$Bisect\_S(.99, .994, .03) = 0.9921$

FIGURE 5.8C
Confidence interval for survivor function, Weibull distribution.

that will exceed the lower specification limit. The supplier can tell the customer with 95% certainty that random variation in this process will not deliver more than 8.9 pieces per thousand below the LSL.

In the case of the upper specification limit, the survivor function would correspond to the nonconforming fraction itself.

Application of this procedure to the placebo data from Lawless (1982, 175–176) returns a 95% confidence interval of [0.1965,0.5130] for $S(10)$, the fraction of the population to survive more than 10 weeks. This matches the reference's results of [0.197,0.513].

Visual Basic for Applications function CONF_WEIBULL_SURVIVOR uses the maximum likelihood approach. Its arguments are the range of raw data, the maximum allowable shape parameter β, the specification limit Q, and the desired confidence level. For the example shown above, =CONF_WEIBULL_SURVIVOR(A6:D55,12,0.03,0.95) returns 0.99215 as the lower 95% confidence limit for the population fraction that exceeds 0.03 for the Weibull distribution whose fitted parameters are $\beta = 1.9686$ and $\theta = 0.48214$.

Minitab 15 will return the confidence limits for the survivor function. Use the reliability/survival menu and parametric distribution analysis with right-censoring (even though no censored results are involved). In the Estimate menu, enter the Q values and also make sure that "log likelihood" is selected as the estimation method. Table 5.2 shows the key results.

Note that Minitab's results for the distribution's shape and scale parameters match the ones from the MathCAD calculation, although Minitab returns 0.9920 instead of 0.99215 for the lower 95% confidence interval for the survivor function at 0.03. =CONF_WEIBULL_SURVIVOR(A6:D55,12,1.3,0.95) returns 0.03361 versus Minitab's 0.03336 for the upper confidence limit for the survivor function, or in this case, the nonconforming fraction above the upper control limit of 1.3, for the same dataset.

The reason for the discrepancy appears to be that, per Minitab's documentation, it uses a normal approximation to find the confidence limits even though the point estimates for the parameters are maximum likelihood estimates. Lawless (1982, 178) states that, unless sample sizes are extremely large—as should be the case for any statement of process performance indices—the likelihood ratio procedure is superior to normal approximation methods. As there are 200 data in this example, the discrepancy is only in the ten-thousandths place.

CONF_WEIBULL_SURVIVOR also returns 0.19645 for the lower 95% confidence limit for $S(10)$ and 0.5130 for the upper 95% confidence limit for $S(10)$ for the placebo data from Lawless (1982, 175–176), although it was necessary to manually reset the bisection interval in the program body; the program is designed for applications in which the nonconforming fraction will be measured in parts per thousand. Minitab gets [0.18711, 0.50145] for the confidence interval. In this case, the discrepancy is in the hundredths place, and this is quite likely because there are only 21 measurements as opposed to 200.

TABLE 5.2

Minitab Calculation of Weibull Distribution Survivor Function

Distribution Analysis: C1
Variable: C1

Censoring Information Count
Uncensored value 200

Estimation Method: Maximum Likelihood
Distribution: Weibull

Parameter Estimates

Parameter	Estimate	Standard Error	95.0% Normal CI Lower	Upper
Shape	1.96857	0.106672	1.77022	2.18915
Scale	0.482136	0.0183042	0.447562	0.519380

Log-Likelihood = 26.907

Goodness-of-Fit
Anderson-Darling (adjusted) = 0.883

Table of Survival Probabilities

Time	Probability	95.0% Normal CI Lower	Upper
0.03	0.995784	0.991995	0.997782
1.30	0.000870	0.000165	0.003336

In the case of the three-parameter Weibull distribution, it should be possible to simply delete the threshold parameter (as calculated by Minitab, StatGraphics, or FitWeibull3.bas) from the data and the specification limit, and then use one of the procedures shown previously.

This section has shown that it is possible to quote lower confidence levels for *PPL* and *PPU* for a process that follows a Weibull distribution. The next will address the more difficult gamma distribution.

Gamma Distribution; Confidence Interval for Nonconforming Fraction

Lawless (1982, 215–217) states that it is difficult to use the likelihood ratio method to get confidence intervals for the gamma distribution's survivor function because of the lack of a closed-form expression for the survivor function. This creates analytical barriers to maximization of $\ln L(\alpha, \gamma)$ subject to the null hypothesis[4]

$$S(t_0 | \alpha, \gamma) = \frac{1}{\Gamma(\alpha)} \int_{x=\gamma t_0}^{\infty} u^{\alpha-1} e^{-u} \, du = S_0$$

where S is the survivor function. In this case, t_0 for reliability purposes would correspond to a specification limit (Q) for determination of the confidence limit for the process performance index. The conditions of S_0 as the γ_1 confidence limit for the survival function at time (or specification limit Q) are as follows:

(1) $\Lambda = -2 \ln L(\tilde{\alpha}, \tilde{\gamma}) + 2 \ln L(\hat{\alpha}, \hat{\gamma}) = \gamma_1$ quantile of χ_1^2

(2) $\tilde{\alpha}$ maximizes $\ln L(\tilde{\alpha}, \tilde{\gamma}(S_0, \tilde{\alpha}))$

(3) $\tilde{\gamma}(S_0, \tilde{\alpha}) = \dfrac{Q_{1-S_0}^*}{Q_0}$ where

$$Q_{1-S_0}^* = 1 - S_0 \text{ quantile of } \frac{1}{\Gamma(\tilde{\alpha})} \int_{\gamma Q_0}^{\infty} u^{\alpha-1} e^{-u} \, du \qquad \text{(Set 5.16)}$$

As with the Weibull distribution, the equation for the shape parameter is in fact redundant, and there are two independent equations with two unknowns: S_0 and $\tilde{\alpha}$.

The reference then recommends maximization of $\ln L(\alpha, \gamma(\alpha))$ by means of numerical methods that do not rely on analytical first derivatives.

$$f'(x) \approx \frac{f(x+\Delta) - f(x)}{\Delta}$$

for any function where Δ is a tiny increment, but tests show that

$$\frac{d \ln L(\tilde{\alpha}, \tilde{\gamma}(S_0, \tilde{\alpha}))}{d\tilde{\alpha}} \approx \frac{\ln L(\tilde{\alpha} + \Delta, \tilde{\gamma}(S_0, \tilde{\alpha} + \Delta)) - \ln L(\tilde{\alpha}, \tilde{\gamma}(S_0, \tilde{\alpha}))}{\Delta}$$

behaves erratically in the region of the solution for $\tilde{\alpha}$. A one-dimensional response surface approach that finds the log likelihood at three points—left, center, and right—and seeks the maximum produced better results.

The argument $\gamma(\alpha)$, or scale parameter as a function of the shape parameter, means that the shape parameter is given and the scale parameter must then satisfy the requirement $S(t_0|\alpha, \gamma(\alpha)) = S_0$. To do this, find $t_{1-S_0}^*$, or $1 - S_0$ quantile of the one-parameter gamma distribution

$$\frac{1}{\Gamma(\alpha)} \int_{x=\gamma t_0}^{\infty} u^{\alpha-1} e^{-u} \, du,$$

and then

$$\gamma(\alpha) = \frac{t_{1-S_0}^*}{t_0}.$$

The procedure then becomes similar to that for the Weibull distribution:

1. Select values for S_0, the candidate value for the confidence limit for the surviving population fraction at specification limit Q_0. These may be the lower and upper ends of a bisection interval.

2. Calculate $\tilde{\alpha}$ and $\tilde{\gamma}$ as follows for each S_0. The goal is to find $\tilde{\alpha}$ to maximize $\ln L(\tilde{\alpha}, \tilde{\gamma}(\tilde{\alpha}))$, so the optimization routine will use $\tilde{\alpha}$ as the variable.

   - Given $\tilde{\alpha}$, compute $Q^*_{1-S_0}$, or the $1 - S_0$ quantile of

$$\frac{1}{\Gamma(\tilde{\alpha})} \int_{\gamma Q_0}^{\infty} u^{\alpha-1} e^{-u} \, du.$$

   Note that the latter is simply GAMMAINV($1-S_0$, $\tilde{\alpha}$,1) in Excel.

   - $\tilde{\gamma}(S_0, \tilde{\alpha}) = \dfrac{Q^*_{1-S_0}}{Q_0}$

   - Compute $\ln L(\tilde{\alpha}, \tilde{\gamma}(\tilde{\alpha}, S_0))$, and iterate the shape parameter until maximized.

3. $\Lambda = 2 \ln L(\hat{\alpha}, \hat{\gamma}) - 2 \ln L(\tilde{\alpha}, \tilde{\gamma})$ and compare to χ^2 for the desired confidence limit. Iterate S_0 (e.g., with bisection) until $\Lambda = \chi^2$.

Lawless (1982, 218) provides the only example that we know of in which an attempt is made to find confidence intervals for the survivor function of a gamma distribution, and one with right-censored data in the bargain. The reference finds point estimates $\hat{\alpha} = 1.625$ and $\hat{\gamma} = 0.08090$ (it reports $1/\gamma = 12.361$) for the dataset itself. The 0.95 confidence interval for $S(15)$, or the survivor function at $Q = 15$, is [0.40,0.66]. The point estimate of $S(15)$ is 0.534.

A VBA response surface algorithm returns $\alpha = 1.642$ and $\gamma = 0.080676$ with a log likelihood of -123.1054 and a point estimate of $S(15) = 0.5349$. StatGraphics returns $\alpha = 1.624$ and $\gamma = 0.080686$. (These numbers contain far more significant figures than the two-digit data justify. They are reported only to allow testing of the solution for conformance to the requirements.) The quantile-quantile plot suggests a poor fit for the gamma distribution, but this does not affect the comparison of the algorithms in question; the issue is whether the algorithm can reproduce the results from Lawless, which it does.

Response surface optimization of the interval [0.535,0.7] by the VBA program yields $\tilde{\alpha} = 1.89383$, $\tilde{\gamma} = 0.0740749$, population (tolerance interval) = 0.6630 at the upper end, and a log likelihood function of -125.0261. Two tests should be performed to ensure that the solution satisfies the conditions of Equation (Set 5.16).

1. $2 \times 125.0261 - 2 \times 123.1054 = 3.8414$ (vs. target of 3.8415).

2. A graph of $\ln L(\tilde{\alpha}, \tilde{\gamma}(S_0, \tilde{\alpha}))$ versus $\tilde{\alpha}$ shows that $\tilde{\alpha}$ maximizes the log likelihood under these conditions.

Bisection of [0.3,0.535] yields $\tilde{\alpha} = 1.36719$, $\tilde{\gamma} = 0.0877359$, and population (tolerance interval) = 0.4033 with a log likelihood of -125.0261, which again satisfies the first requirement. A plot of $\ln L(\tilde{\alpha}, \tilde{\gamma}(S_0, \tilde{\alpha}))$ versus $\tilde{\gamma}$ indicates that the maximum log likelihood is at about $\tilde{\alpha} = 1.36719$.

Now consider 50 data that were simulated from a gamma distribution with $\alpha = 3$ and $\gamma = 12$, and let the critical quality characteristic be parts per million of an impurity or contaminant. The upper specification limit is 1.0 ppm. What is the lower 97.5% confidence interval for the process performance index *PPU*? The simulated data are in SIMULAT.XLS under tab Gamma_a3_g12.

=Conf_Gamma_Survivor(B6:B55,1,1,0.95) finds $\hat{\alpha} = 2.47535$, $\hat{\gamma} = 10.2879$, and log likelihood 30.37896. (StatGraphics gets 2.47451 and 10.2844 for the shape and scale parameters, respectively.) Bisection of the interval defined by the point estimate 0.000933 and initial x_R 0.2 yields $\tilde{\alpha} = 1.88599$, $\tilde{\gamma} = 6.77566$, and 97.5% upper confidence limit of 0.0073332 for the nonconforming fraction above 1.0 ppm. The log likelihood is 28.4582 under these conditions. The 97.5% lower confidence limit for PPU (which corresponds to the 97.5% upper confidence limit for the nonconforming fraction) is $\frac{1}{3}\Phi^{-1}(0.0073332) = 0.814$.

The next step is to check the solution, and $2 \times (30.37896 - 28.4582) = 3.8415$ shows that the first requirement has been met. Figure 5.9 shows that $\tilde{\alpha} = 1.88599$ meets the second requirement as well. The log likelihood function $\ln L(\alpha, \gamma)$ is designed to accommodate right-censoring for the example from Lawless, but this dataset is uncensored, and the denominator of γ_tilde_function is the specification limit of 1.0 ppm. It also illustrates the one-dimensional response surface algorithm that optimizes $\tilde{\alpha}$.

Case 2 is placed to the left of Case 1 because it would almost certainly arise after the situation shown at the right as opposed to before it. In Case 1, the left point yields the highest log likelihood, so it becomes the center point for the next iteration; this moves the bracketed interval toward Case 2. In Case 2, the center point yields the highest log likelihood, so the next step is to halve the size of the interval. This approach yielded better results than efforts to drive the first derivative of the log likelihood function to zero.

In summary, it was possible to write a VBA function (Conf_Gamma_Survivor.bas) that reproduces the results from Lawless (1982, 218), and for which the results fulfill the conditions that: (1) the likelihood ratio $\Lambda = 2 \ln L(\hat{\alpha}, \hat{\gamma}) - 2 \ln L(\tilde{\alpha}, \tilde{\gamma})$ equals the χ^2 target statistic, and (2) $\tilde{\alpha}$ maximizes the log likelihood function for S_0 at the confidence limit. Application of the same procedure to simulated data also delivers results that meet the necessary conditions. *This suggests that it is in fact possible to find confidence limits for the process performance index of a gamma distribution*, and the procedure could conceivably be extended to reliability and actuarial applications as well. As it was possible to test the algorithm against only the example from Lawless, any results that are given to a customer should have a caveat, but it at least shows that a solution to this challenging problem might, in fact, exist.

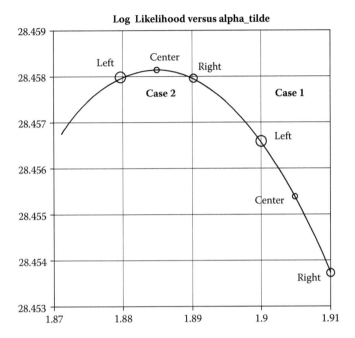

$$\begin{bmatrix} S_0 \\ \alpha\_tilde \\ \gamma\_tilde \end{bmatrix} = \begin{bmatrix} 7.3332.10^{-3} \\ 1.88599 \\ 6.77566 \end{bmatrix}$$

$\ln L(\alpha\_tilde, \gamma\_tilde) = 28.45819$ This is the maximum per the next two lines

$\ln L\,(\alpha\_tilde - 0.0001, \gamma\_tilde\_function(S_0, \alpha\_tilde - 0.0001)) = 28.45814$

$\ln L\,(\alpha\_tilde + 0.0001, \gamma\_tilde\_function(S_0, \alpha\_tilde + 0.0001)) = 28.45814$

Where $Q\_star(S, \alpha) := qgamma(1 - S_0, \alpha)$ $\gamma\_tilde\_function(S_0, \alpha) := \dfrac{Q\_star(1 - S_0, \alpha)}{1}$

Log Likelihood function: the conditional test x<D excludes the right-censored observations

$$\ln L(\alpha, \gamma) := r \cdot \alpha \cdot \ln(\gamma) - r.\ln(\Gamma(\alpha)) + \left| \sum_{i=1}^{n} \left[(\alpha - 1) \cdot \ln(x_i) \cdot (x_i < D) - \gamma \cdot x_i \cdot (xi < D) \right] \right. \cdots$$
$$+ (n - r) \cdot \ln \left| 1 - \int_0^{iD} \frac{\gamma^{\alpha}}{\Gamma(\alpha)} \cdot y^{\alpha - 1} \cdot \exp(-\gamma \cdot y) \, dy \right|$$

FIGURE 5.9
versus $\tilde{\alpha}$, upper confidence limit.

Exercise

Exercise 5.1

A normally distributed process has an upper specification limit of 800 microns. Fifty pieces have a grand average of 788.70 microns and a standard deviation of 3.0 microns. What are the point estimate and the 95% lower confidence limit for *PPU*?

Solution

Exercise 5.1

$$\hat{PPU} = \frac{USL - \hat{\mu}}{3\hat{\sigma}} = \frac{(800 - 788.70) \text{ microns}}{(3 \times 3) \text{ microns}} = 1.256$$

There are 3.766 standard deviations between the grand average and the *USL*, which means $k = 3.766$ is the tolerance factor. Juran and Gryna (1988, AII.36) or an equivalent table shows 3.766 to be the one-sided tolerance factor for 99.9% of the population with 95% confidence. $\Phi^{-1}(0.999) = 3.090$ so there is 95% confidence that *PPU* is no less than $3.090 \div 3 = 1.03$. This means there is 95% confidence that no more than 1 piece per thousand will exceed the upper specification.

Endnotes

1. The reference uses K_α and K_γ for the standard normal deviates for the population and confidence level, respectively. This can be somewhat confusing because it also uses K (no subscript) for the tolerance factor.
2. See also Chou et al (1990), Owen (1965), and Papp (1992) for the background equations.
3. See also N. F. Zhang, G. A. Stenback, and G. S. Wasserman, "Interval Estimation of Process Capability Index C_{pk}," *Commun. Statist. Theory and Methods* 19, 4455–4470.
4. Lawless (p. 207) uses $k = \alpha$ and $\alpha = 1/\gamma$, and defines

$$Q(k, x) = \frac{1}{\Gamma(k)} \int_x^\infty u^{k-1} e^{-u} \, du \quad \text{where } x = \frac{t}{\alpha}.$$

6

The Effect of Gage Capability

The book has assumed so far that all measurements are exact. As an example, 15.83 microns means that the product feature is exactly 15.83 microns wide, with uncertainty only in the last figure. The unfortunate truth is that gages and instruments are themselves often the source of variation, and this variation affects

1. Estimates of process capability and process performance
2. Power and average run lengths for control charts
3. Outgoing quality, noting that borderline nonconforming pieces may be accepted while borderline conforming pieces are rejected. Guardbanding, or placement of rejection limits inside the specification limits, will protect the customer but it will also result in the loss of conforming parts.

In the previous example, 0.01 microns, or 10 nanometers, is the smallest readable dimension on the instrument or gage. This is known as the *resolution*, and the Automotive Industry Action Group (AIAG, 2002, 13) says that it must divide the tolerance into 10 or more parts (Rule of Tens, or 10 to 1 rule). This is an absolute minimum requirement, and such a measurement system would resemble an attribute or nonparametric system with only 10 possible categories more than a genuine continuous scale measurement. The resolution's relevance, or lack thereof, to the actual quality characteristic depends entirely on the gage's accuracy and variation.

Gage Accuracy and Variation

AIAG (2002, 6–8) summarizes potential sources of gage accuracy and variation and adds (pp. 13–14) the SWIPE acronym. SWIPE (an easily-remembered contraction of SWIPPE, or standard, workpiece, instrument, person, procedure, environment) provides six categories for a cause-and-effect or fishbone diagram for the measurement system:

1. Standard, against which the gage is calibrated
2. Workpiece, the part that is being measured

3. Instrument (or gage)
- This is the source of *repeatability*, or variations in measurements made on a single part by the same inspector

4. Person
- The inspector or operator, in combination with the design of the gage and/or workpiece, is the source of *reproducibility*, or the ability of different inspectors to get the same measurement from a single part.
- If it is possible for two inspectors to read a dial gage differently, or align the crosshairs of a digital micrometer differently, there will almost certainly be reproducibility variation. All inspectors will, however, read the same results from a digital display.
- Conformance to the work instruction (item 5) does not eliminate reproducibility variation. If the procedure says to align the digital micrometer's crosshairs on a product feature, but the feature is not sufficiently distinct to eliminate any question as to its exact location, different people will do it differently.

5. Procedure, for example, the work instruction for the measurement

6. Environment
- The impact of environment, which generally includes temperature and humidity but may include other factors as well, cannot be overemphasized. Ford and Crowther (1926, 85) cited the influence of radiated body heat on the dimensions of microinch-step Johansson ("Jo") blocks and added (1930, 207–208): "Arguments between a mechanic on the warm side of a shop with one on the cold side often became legal battles between seller and buyer which involved broken contracts, months or even years of litigation, and often not inconsiderable damage awards by courts."

As with the process itself, the gage may be accurate and/or capable. An accurate gage returns, on average, the dimension of the standard against which it is calibrated. A capable gage returns the same measurement consistently, and this is where gage reproducibility and repeatability become important. The AIAG reference lists the following influences on accuracy:

1. Accuracy, or proximity to the true value
2. Bias, or a systematic difference between the gage's output and a reference value such as a standard
3. Stability or drift, which refers to a change in bias over time
4. Linearity, or a change in bias that depends on the magnitude of the measurement

Further discussions will assume that gages and instruments have been calibrated, which is required by the International Organization for Standardization (ISO 9001:2008). Calibration does not, however, address the capability of the gage.

Gage Capability

Gage capability is measured directly in terms of gage reproducibility and repeatability (GRR), Percent Tolerance Consumed by (gage) Capability or PTCC, or precision-to-tolerance (P/T) ratio. The latter two are the same thing, while the former is the gage standard deviation in the same units of measurement as the quality characteristic.

$$GRR = \sigma_{gage} = \sqrt{\sigma^2_{repeatability} + \sigma^2_{reproducibility}} \quad \text{(AIAG, 2002, p. 114)}$$

where

Equipment variation $= \sigma_{repeatability}$ and

Evaluator variation $= \sigma_{reproducibility}$

$$PTCC = \frac{c \times \sigma_{gage}}{USL - LSL} \times 100\% \quad \text{where } c = 5.15 \text{ or } c = 6 \text{ depending on the reference}$$

In Barrentine (1991), $c = 5.15$, while $c = 6$ in Montgomery (1991, 396) and Hradesky, (1995, 420–421). The explanation for these respective factors is that 99% of a normal distribution is contained within 5.15 standard deviations of its mean (2.575 standard deviations to either side of the mean), while six standard deviations make up the "process width." A gage is generally considered adequate if the P/T ratio is 10% or less, marginal if the P/T ratio is between 10 and 30%, and not capable if the P/T exceeds 30%. Barrentine (1991) and the Automotive Industry Action Group (2002) show how to perform reproducibility and repeatability (R&R) studies. Minitab (Stat/Quality Tools/Gage Study menu) and StatGraphics (SPC/Gage Studies) can handle the necessary calculations.

Suppression of Repeatability Variation

It is possible, although at the cost of more work, to suppress repeatability variation by using the average of n measurements of the quality characteristic in question. Then the total gage variation is

$$\sigma^2_{gage} = \sigma^2_{reproducibility} + \frac{1}{n}\sigma^2_{repeatability}$$

If the gage or instrument is highly automated and is not the bottleneck or capacity-constraining operation, there is little reason not to do this if repeatability is a significant part of the gage variation. If the measurement is labor intensive, this procedure might be invoked for parts that a single measurement places marginally inside or outside the specification limit.

Now that this chapter has identified the issue of gage capability, the next section will treat its effects on statistical process control.

Gage Capability and Statistical Process Control

The control limits for process average and variation depend on the overall estimates for process variation, which incorporates the process and gage variation: $\sigma_{total}^2 = \sigma_{process}^2 + \sigma_{gage}^2$. Consider a theoretical (mean and standard deviation given) control chart in which

$$UCL = \mu_0 + 3\frac{\sigma_{total}}{\sqrt{n}}.$$

Now determine the chance that a sample average exceeds the *UCL* if the process mean shifts by δ *process* standard deviations.

$$\Pr(\bar{x} > UCL) = 1 - \Phi\left(\frac{UCL - (\mu_0 + \delta\sigma_{process})}{\frac{\sigma_{total}}{\sqrt{n}}}\right) = 1 - \Phi\left(\frac{3\frac{\sigma_{total}}{\sqrt{n}} - \delta\sigma_{process}}{\frac{\sigma_{total}}{\sqrt{n}}}\right)$$

This leads to Equation (6.1) for the power of an x-bar chart to detect a shift of δ process standard deviations when variation from the gage is present.

$$\gamma(\delta, n) = 1 - \Phi\left(3 - \frac{\delta\sigma_{process}\sqrt{n}}{\sqrt{\sigma_{process}^2 + \sigma_{gage}^2}}\right) \tag{6.1}$$

Comparison to Equation (1.3) shows that this chart's power is lower than one whose gage variation is zero.

$$\gamma(\delta, n) = 1 - \Phi\left(\frac{\left(\mu_0 + 3\frac{\sigma}{\sqrt{n}}\right) - (\mu_0 + \delta\sigma)}{\frac{\sigma}{\sqrt{n}}}\right) = 1 - \Phi(3 - \delta\sqrt{n}) \tag{1.3}$$

As shown in Chapter 3, the power of the s chart to detect an increase in process variation is

$$\gamma = 1 - F\left(\frac{(n-1)UCL^2}{\sigma_{total}^2}\right)$$

TABLE 6.1

Effect of Gage Capability on Power of the s Chart

Case 1: $\sigma_{total} = \sigma_{process} = 4$ mils	Case 2: $\sigma_{process} = 4$ mils and $\sigma_{gage} = 3$ mils σ_{total} is therefore 5 mils
For a false alarm risk of 0.00135, χ^2 with 3 degrees of freedom is 15.630.	
$s^2_{UCL} = \chi^2 \dfrac{\sigma_0^2}{(n-1)}$	$s^2_{UCL} = 15.630 \dfrac{25(mils)^2}{3} = 130.25(mils)^2$
$= 15.630 \dfrac{16(mils)^2}{3} = 83.36(mils)^2$	
Now let $\sigma_{total} = \sigma_{process} = 5$ mils	Now let $\sigma_{process} = 5$ mils
$\gamma = 1 - F\left(\dfrac{(n-1)UCL^2}{\sigma_{total}^2} \right)$	$\gamma = 1 - F\left(\dfrac{(n-1)UCL^2}{5^2 + 3^2} \right)$
$= 1 - F\left(\dfrac{3 \times 83.36}{5^2} \right) = 0.0185$	$= 1 - F\left(\dfrac{3 \times 130.25}{34} \right) = 0.00994$
or =CHIDIST(83.36*3/25,3) in Excel	or =CHIDIST(130.25*3/34,3) in Excel

where F is the cumulative distribution function for the chi square statistic with $n - 1$ degrees of freedom. Now consider the two cases in Table 6.1, given a sample size of 4. For a false alarm risk of 0.00135 at the upper control limit,

$$\chi^2 = \frac{(n-1)s^2}{\sigma_0^2} \Rightarrow s^2_{UCL} = \chi^2 \frac{\sigma_0^2}{(n-1)}.$$

This example shows that *gage variation reduces the ability of the s chart to detect an increase*, in this case from 4 mils to 5 mils, *in the process's standard variation*. The same issue will apply to range charts for normal and nonnormal variation. Gage capability will also affect the estimates for process capability and process performance.

Gage Capability and Process Capability

The total variance of any measurement is $\sigma_{process}^2 + \sigma_{gage}^2$, so the observed process performance index is less than the actual process performance index.

$$PP_{observed} = \frac{USL - LSL}{\sqrt{\sigma_{process}^2 + \sigma_{gage}^2}} \quad \text{versus} \quad PP_{actual} = \frac{USL - LSL}{\sigma_{process}}$$

Similar reasoning applies to the other capability and performance indices, and AIAG (2002, 191–194) has an appendix on this subject. The good news is that, if the GRR is known—and this is mandatory under most quality system standards—the actual capability or performance index is better than the observed index, and it is easy to determine the actual process standard deviation: $\sigma_{process} = \sqrt{\sigma_{total}^2 - \sigma_{gage}^2}$.

The bad news is that, if GRR is poor, it may be necessary to move product acceptance limits inside the specification limits (sometimes known as *guardbanding*) to avoid the chance of shipping nonconforming parts due to measurement variation in their favor. This will of course result in the rejection of parts that actually meet specification. Levinson (1995, 1996) discusses this issue further.

Gage Capability and Outgoing Quality

Unless the process capability is so exceptional as to preclude any reasonable chance of nonconforming product, inspections and tests will be necessary. It is generally assumed that inspections and tests will reject all nonconforming parts and accept all parts that are within specification. Gage variation, however, makes this assumption a bad one.

Let $f(x)$ be the probability density function, whether normal or nonnormal, for the actual part dimension, and $g(y|x)$ the probability of getting measurement y given actual dimension

$$x.\ g(y|x) = \frac{1}{\sqrt{2\pi}\sigma_{gage}} \exp\left(-\frac{1}{2\sigma_{gage}^2}(y-x)^2\right),$$

which is simply the normal probability density function for a mean of x and a standard deviation σ_{gage}. Abbreviate this as $\phi(y|x)$. Then the joint probability of dimension x and measurement y is $h(x,y) = f(x)\phi(y|x)$ and

1. Nonconforming parts accepted (corresponds to consumer's risk):

$$\int_{USL}^{\infty}\int_{LSL}^{USL} h(x,y)\,dy\,dx + \int_{min}^{LSL}\int_{LSL}^{USL} h(x,y)\,dy\,dx$$

where min is the minimum possible measurement below the lower specification limit (technically, negative infinity for the normal distribution, and zero or the threshold parameter for the gamma and Weibull distributions).

- This is the double integral of the joint probability that the part dimension exceeds the specification *and* the chance that the measurement will be within the specification.

- The double integral is easily simplified as follows:

$$\int_{USL}^{\infty}\int_{LSL}^{USL} h(x,y)\,dy\,dx = \int_{USL}^{\infty} f(x)\int_{LSL}^{USL}\phi(y|x)\,dy\,dx$$

$$\approx \int_{USL}^{\infty} f(x)\int_{-\infty}^{USL}\phi(y|x)\,dy\,dx = \int_{USL}^{\infty} f(x)\Phi\left(\frac{USL-x}{\sigma_{gage}}\right)dx$$

noting that the second term of the exact integral

$$\int_{USL}^{\infty} f(x) \left[\Phi\left(\frac{USL - x}{\sigma_{gage}} \right) - \Phi\left(\frac{LSL - x}{\sigma_{gage}} \right) \right] dx$$

measures the chance that a measurement from a part above the *USL* will be below the *LSL*. This is essentially zero in any practical situation.

- Finally, the consumer's risk of getting nonconforming parts is

$$\int_{USL}^{\infty} f(x) \Phi\left(\frac{USL - x}{\sigma_{gage}} \right) dx + \int_{min}^{LSL} f(x) \Phi\left(\frac{x - LSL}{\sigma_{gage}} \right) dx$$

where the first term is the chance of getting a part above the *USL* but measuring it below the *USL*, and the second is the chance of getting a part below the *LSL* but measuring it above the *LSL*.

2. Conforming parts rejected (corresponds to producer's risk):

$$\int_{LSL}^{USL} f(x) \left[\Phi\left(\frac{LSL - x}{\sigma_{gage}} \right) + \Phi\left(\frac{x - USL}{\sigma_{gage}} \right) \right] dx$$

where the first term is the chance of getting a part within specification but measuring it below the *LSL*, and the second is the chance of measuring it above the *USL*.

3. Conforming parts accepted:

$$\int_{LSL}^{USL} f(x) \left[\Phi\left(\frac{USL - x}{\sigma_{gage}} \right) - \Phi\left(\frac{LSL - x}{\sigma_{gage}} \right) \right] dx$$

where, given a dimension within the specification limits, the first term is the chance that the measurement will be below the *USL* and the second term the chance it will be *below* the *LSL*.

4. Nonconforming parts rejected:

$$\int_{USL}^{\infty} f(x) \Phi\left(\frac{x - USL}{\sigma_{gage}} \right) dx + \int_{min}^{LSL} f(x) \Phi\left(\frac{LSL - x}{\sigma_{gage}} \right) dx$$

Measurement y

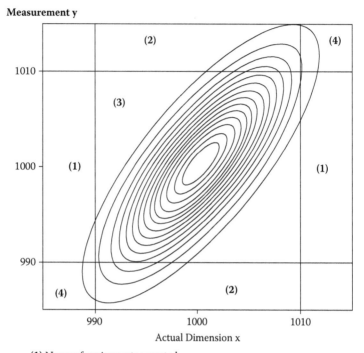

(1) Nonconforming parts accepted
(2) Conforming parts rejected
(3) Conforming parts accepted
(4) Nonconforming parts rejected

FIGURE 6.1
Effect of gage capability on outgoing quality.

where the first term is the chance that a dimension will be above the *USL* and measured as such, and the second term is the chance that a dimension will be below the *LSL* and measured as such.

Figure 6.1 shows a contour plot for a process with specification limits of [990,1010], $\sigma_{\text{process}} = 5$, and $\sigma_{\text{gage}} = 4$. (Such a process is not even marginally capable, but it was necessary to use relatively high standard deviations to get the contour lines to fill the figure.) The numbers in the figure correspond to the equations shown previously.

As a check on these equations, add the consumer's risk of getting nonconforming parts (1) and the chance of rejecting nonconforming parts (4). Note that

$$\Phi\left(\frac{x - USL}{\sigma_{gage}}\right) = 1 - \Phi\left(\frac{USL - x}{\sigma_{gage}}\right)$$

and

$$\Phi\left(\frac{LSL-x}{\sigma_{gage}}\right)=1-\left(\frac{x-LSL}{\sigma_{gage}}\right)$$

$$\int_{USL}^{\infty}f(x)\Phi\left(\frac{USL-x}{\sigma_{gage}}\right)dx+\int_{min}^{LSL}f(x)\Phi\left(\frac{x-LSL}{\sigma_{gage}}\right)dx+$$

$$\cdots\int_{USL}^{\infty}f(x)\Phi\left(\frac{x-USL}{\sigma_{gage}}\right)dx+\int_{min}^{LSL}f(x)\Phi\left(\frac{LSL-x}{\sigma_{gage}}\right)dx$$

$$=\int_{USL}^{\infty}f(x)\left[\Phi\left(\frac{USL-x}{\sigma_{gage}}\right)+1-\Phi\left(\frac{USL-x}{\sigma_{gage}}\right)\right]dx$$

$$+\int_{min}^{LSL}f(x)\left[\Phi\left(\frac{x-LSL}{\sigma_{gage}}\right)+1-\Phi\left(\frac{x-USL}{\sigma_{gage}}\right)\right]dx$$

$$=\int_{USL}^{\infty}f(x)dx+\int_{min}^{LSL}f(x)dx$$

The result is the total fraction of nonconforming parts. Now add the chance of rejecting good parts (2) and the chance of accepting good parts (3).

$$\int_{LSL}^{USL}f(x)\left[\Phi\left(\frac{LSL-x}{\sigma_{gage}}\right)+\Phi\left(\frac{x-USL}{\sigma_{gage}}\right)\right]dx$$

$$+\int_{LSL}^{USL}f(x)\left[\Phi\left(\frac{USL-x}{\sigma_{gage}}\right)-\Phi\left(\frac{LSL-x}{\sigma_{gage}}\right)\right]dx$$

$$=\int_{LSL}^{USL}f(x)\left[\Phi\left(\frac{LSL-x}{\sigma_{gage}}\right)+\Phi\left(\frac{x-USL}{\sigma_{gage}}\right)\right]dx$$

$$+\int_{LSL}^{USL}f(x)\left[1-\Phi\left(\frac{x-USL}{\sigma_{gage}}\right)-\Phi\left(\frac{LSL-x}{\sigma_{gage}}\right)\right]dx$$

$$=\int_{LSL}^{USL}f(x)dx \quad \text{and then the total is} \quad \int_{USL}^{\infty}f(x)dx+\int_{min}^{LSL}f(x)dx$$

$$+\int_{LSL}^{USL}f(x)dx=\int_{min}^{\infty}f(x)dx=1$$

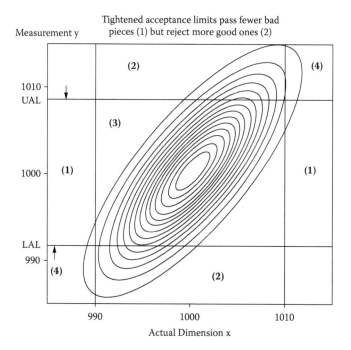

FIGURE 6.2
Effect of guardbanding on producer's and consumer's risks.

Guardbanding, where the upper acceptance limit is $UAL < USL$ and lower acceptance limit is $LAL > LSL$, reduces the chance of accepting nonconforming work, but it also increases the producer's risk as shown in Figure 6.2.

Equation (Set 6.2) summarizes the results: outgoing quality as a function of acceptance limits.

$$\int_{USL}^{\infty} f(x)\Phi\left(\frac{UAL-x}{\sigma_{gage}}\right)dx + \int_{min}^{LSL} f(x)\Phi\left(\frac{x-LAL}{\sigma_{gage}}\right)dx \quad \text{Bad parts accepted}$$

$$\int_{LSL}^{USL} f(x)\left(\Phi\left(\frac{LAL-x}{\sigma_{gage}}\right)+\Phi\left(\frac{x-UAL}{\sigma_{gage}}\right)\right)dx \quad \text{Good parts rejected}$$

$$\int_{LSL}^{USL} f(x)\left(\Phi\left(\frac{UAL-x}{\sigma_{gage}}\right)-\Phi\left(\frac{LAL-x}{\sigma_{gage}}\right)\right)dx \quad \text{Good parts accepted}$$

$$\int_{USL}^{\infty} f(x)\Phi\left(\frac{x-UAL}{\sigma_{gage}}\right)dx + \int_{min}^{LSL} f(x)\Phi\left(\frac{LAL-x}{\sigma_{gage}}\right)dx \quad \text{Bad parts rejected}$$

(Set 6.2)

Gage capability and outgoing quality, normal distribution

$$
\begin{bmatrix}
\text{USL} \\
\text{LSL} \\
\text{UAL} \\
\text{LAL} \\
\text{mean} \\
s_{process} \\
s_{gage} \\
\text{min} \\
\infty
\end{bmatrix}
:=
\begin{bmatrix}
1010 \\
990 \\
1010 \\
990 \\
1000 \\
5 \\
4 \\
950 \\
1050
\end{bmatrix}
$$

Bad parts accepted

$$
\int_{USL}^{\infty} dnorm(x, mean, s_{process}) \cdot cnorm\left(\frac{UAL-x}{s_{gage}}\right) dx \ldots = 0.01533
$$

$$
+ \int_{min}^{LSL} dnorm(x, mean, s_{process}) \cdot cnorm\left(\frac{x-LAL}{s_{gage}}\right) dx
$$

Good parts rejected

$$
\int_{LSL}^{USL} dnorm(x, mean, s_{process}) \cdot \left(cnorm\left(\frac{LAL-x}{s_{gage}}\right) + cnorm\left(\frac{x-UAL}{s_{gage}}\right) \right) dx = 0.08818
$$

Good parts accepted

$$
\int_{LSL}^{USL} dnorm(x, mean, s_{process}) \cdot \left(cnorm\left(\frac{UAL-x}{s_{gage}}\right) - cnorm\left(\frac{LAL-x}{s_{gage}}\right) \right) dx = 0.86632
$$

Bad parts rejected

$$
\int_{USL}^{\infty} dnorm(x, mean, s_{process}) \cdot cnorm\left(\frac{x-UAL}{s_{gage}}\right) dx \ldots = 0.03017
$$

$$
+ \int_{min}^{LSL} dnorm(x, mean, s_{process}) \cdot cnorm\left(\frac{LAL-x}{s_{gage}}\right) dx
$$

FIGURE 6.3
Calculation of outgoing quality given gage and process variation.

Figure 6.3 shows a MathCAD spreadsheet that performs these calculations for the example with specification limits of [990,1010], $\sigma_{process} = 5$, and $\sigma_{gage} = 4$. Note that the four quantities involved, 0.01533, 0.08818, 0.86632, and 0.03017, add up to 1.0. (dnorm is the normal distribution's probability density function, and cnorm is the cumulative standard normal distribution $\Phi(z)$.)

The Visual Basic for Applications function Gage_Quality, which reads the integration range [a,b] and the acceptance limit directly from the spreadsheet (select these three cells as the function's argument), uses Romberg integration with a convergence tolerance of one millionth to get the same results as shown in Table 6.2. The sign of the acceptance limit tells the function whether the numerator of the cumulative standard normal distribution should be the acceptance limit minus x (positive acceptance limit) or x minus the acceptance

TABLE 6.2

Outgoing Quality, Romberg Integration Method

	Integration Limits		Acceptance		
	a	b	Limit		
Bad parts accepted above UAL	1010	1040	1010	0.007665358	
Bad parts accepted below LAL	960	990	−990	0.007665358	0.01533
Good parts rejected above UAL	990	1010	−1010	0.044090141	
Good parts rejected below LAL	990	1010	990	0.044090141	0.08818
Good parts accepted below UAL	990	1010	1010	0.910409595	
minus rejected below LAL	990	1010	990	0.044090141	0.86632
Bad parts rejected above UAL	1010	1040	−1010	0.015084775	
Bad parts rejected below LAL	960	990	990	0.015084775	0.03017
					1.00000

limit (negative acceptance limit). In both cases, the user can tighten the acceptance limits and see the resulting effect on the outgoing quality.

Exercises

Exercise 6.1

A process that is known to follow the normal distribution has an observed standard deviation of 5 mils and an observed process capability index C_p of 1.2. If the standard deviation of the gage is 3 mils, what is the actual process capability?

Exercise 6.2

Given a process with specification limits [990, 1010] microns, gage reproducibility of 0.3 microns, and gage repeatability of 0.4 microns, if the percent tolerance consumed by gage capability is defined as

$$PTCC = \frac{6 \times \sigma_{gage}}{USL - LSL} \times 100\%,$$

then:

1. What is the PTCC or P/T ratio?
2. If the customer requires a P/T ratio of 10% or less, how could this be achieved?

Exercise 6.3

A process has specification limits [980, 1020] microns, process standard deviation of 5 microns, and a gage standard deviation of 3 microns.

1. How much nonconforming product will be shipped to the customer?
2. What must the acceptance limits be to make sure that no more than 10 nonconforming pieces per million will reach the customer (assuming no process shift)? If so, how much conforming product will be rejected?

Use of the Gage_Quality Visual Basic for Applications (VBA) function to solve this will require reprogramming of the joint probability density function as follows. There is no change in the mean or s_process (process standard deviation) but s_gage must be 3. The reason for the hard coding is to allow accommodation of the gamma and Weibull distributions.

Function $f(x, \text{acceptance\_limit})$ As Double

Mean $= 1000$

s_process $= 5$

s_gage $= 3$

Solutions

Exercise 6.1

$$C_{p,observed} = \frac{USL - LSL}{6\sqrt{\sigma^2_{process} + \sigma^2_{gage}}} = 1.2 \quad \text{and} \quad \sqrt{\sigma^2_{process} + \sigma^2_{gage}} = 5 \text{ mils}$$

Also $\sigma_{gage} = 3$ mils $\Rightarrow \sigma_{process} = 4$ mils

$$C_{p,actual} = C_{p,observed} \times \frac{\sqrt{\sigma^2_{process} + \sigma^2_{gage}}}{\sigma_{process}} = 1.2 \times \frac{5 \text{ mils}}{4 \text{ mils}} = 1.5$$

Exercise 6.2

1. $PTCC = \dfrac{6 \times \sqrt{0.3^2 + 0.4^2} \ \text{micons}}{(1010 - 990) \ \text{microns}} \times 100\% = 15\%$

2. Take the average of n replicate measurements. n must be 8 or more.

$$PTCC = \frac{6 \times \sqrt{0.3^2 + \frac{0.4^2}{n}} \ \text{micons}}{(1010 - 990) \ \text{microns}} \times 100\% \le 10\%$$

$$\Rightarrow 6 \times \sqrt{0.3^2 + \frac{0.4^2}{n}} \le 2 \quad \Rightarrow 0.3^2 + \frac{0.4^2}{n} \le \frac{4}{36} \quad \Rightarrow n \ge 8$$

Exercise 6.3

1. Nonconforming product shipped to the customer: 23 pieces per million

Gage capability and outgoing quality, normal distribution

$$
\begin{bmatrix} \text{USL} \\ \text{LSL} \\ \text{UAL} \\ \text{LAL} \\ \text{mean} \\ s_{process} \\ s_{gage} \\ \text{min} \\ \infty \end{bmatrix} := \begin{bmatrix} 1010 \\ 980 \\ 1020 \\ 980 \\ 1000 \\ 5 \\ 3 \\ 950 \\ 1050 \end{bmatrix}
$$

Bad parts accepted

$$
\int_{USL}^{\infty} \text{dnorm}\langle x, \text{mean}, s_{process}\rangle \cdot \text{cnorm}\left(\frac{\text{UAL}-x}{s_{gage}}\right) dx \dots = 2.3089 \cdot 10^{-5}
$$

$$
+ \int_{\min}^{LSL} \text{dnorm}\langle x, \text{mean}, s_{process}\rangle \cdot \text{cnorm}\left(\frac{x-\text{LAL}}{s_{gage}}\right) dx
$$

Good parts rejected

$$
\int_{LSL}^{USL} \text{dnorm}\langle x, \text{mean}, s_{process}\rangle \cdot \left(\text{cnorm}\left(\frac{\text{LAL}-x}{s_{gage}}\right) + \text{cnorm}\left(\frac{x-\text{UAL}}{s_{gage}}\right)\right) dx = 5.63391 \cdot 10^{-4}
$$

Good parts accepted

$$
\int_{LSL}^{USL} \text{dnorm}\langle x, \text{mean}, s_{process}\rangle \cdot \left(\text{cnorm}\left(\frac{\text{UAL}-x}{s_{gage}}\right) - \text{cnorm}\left(\frac{\text{LAL}-x}{s_{gage}}\right)\right) dx = 0.99937
$$

Bad parts rejected

$$
\int_{USL}^{\infty} \text{dnorm}\langle x, \text{mean}, s_{process}\rangle \cdot \text{cnorm}\left(\frac{x-\text{UAL}}{s_{gage}}\right) dx \dots = 4.02535 \cdot 10^{-5}
$$

$$
+ \int_{\min}^{LSL} \text{dnorm}\langle x, \text{mean}, s_{process}\rangle \cdot \text{cnorm}\left(\frac{\text{LAL}-x}{s_{gage}}\right) dx
$$

or, using the VBA function Gage_Quality, whose arguments are the cells that contain the integration limits and acceptance limit,

	Integration Limits		Acceptance		
	a	b	Limit		
Bad parts accepted above UAL	1020	1040	1020	1.15445E-05	
Bad parts accepted below LAL	960	980	−980	1.15445E-05	0.000023
Good parts rejected above UAL	980	1020	−1020	0.000281698	
Good parts rejected below LAL	980	1020	980	0.000281698	0.000563

	Integration Limits		Acceptance		
	a	b	Limit		
Good parts accepted below UAL	980	1020	1020	0.99965496	
minus rejected below LAL	980	1020	980	0.000281698	0.999373
Bad parts rejected above UAL	1020	1040	−1020	2.01267E-05	
Bad parts rejected below LAL	960	980	980	2.01267E-05	0.000040
					1.000000

2. Adjust the acceptance limits to reduce the nonconforming pieces to 10 per million. The target can be bracketed as if with bisection. Limits of [982,1018] yield 10.3 ppm, which suggests that the limits should be a little wider than [983,1017]; these yield 6.1 ppm. [982.5,1017.5] yields 8 ppm, and so on until [982.06,1017.94] yields 10.00 ppm. At this point, however, two good pieces out of every thousand will be rejected.

Gage capability and outgoing quality, normal distribution

$$
\begin{bmatrix}
\text{USL} \\
\text{LSL} \\
\text{UAL} \\
\text{LAL} \\
\text{mean} \\
s_{process} \\
s_{gage} \\
\text{min} \\
\infty
\end{bmatrix}
:=
\begin{bmatrix}
1020 \\
980 \\
1017.94 \\
982.06 \\
1000 \\
5 \\
3 \\
950 \\
1050
\end{bmatrix}
$$

Bad parts accepted

$$
\int_{USL}^{\infty} \text{dnorm}\langle x, \text{mean}, s_{process}\rangle \cdot \text{cnorm}\left(\frac{UAL-x}{s_{gage}}\right) dx \ldots = 9.99853 \cdot 10^{-6}
$$

$$
+ \int_{min}^{LSL} \text{dnorm}\langle x, \text{mean}, s_{process}\rangle \cdot \text{cnorm}\left(\frac{x-LAL}{s_{gage}}\right) dx
$$

Good parts rejected

$$
\int_{LSL}^{USL} \text{dnorm}\langle x, \text{mean}, s_{process}\rangle \cdot \left(\text{cnorm}\left(\frac{LAL-x}{s_{gage}}\right) + \text{cnorm}\left(\frac{x-UAL}{s_{gage}}\right)\right) dx = 2.03982 \cdot 10^{-3}
$$

Good parts accepted

$$
\int_{LSL}^{USL} \text{dnorm}\langle x, \text{mean}, s_{process}\rangle \cdot \left(\text{cnorm}\left(\frac{UAL-x}{s_{gage}}\right) - \text{cnorm}\left(\frac{LAL-x}{s_{gage}}\right)\right) dx = 0.9979
$$

Bad parts rejected

$$
\int_{USL}^{\infty} \text{dnorm}\langle x, \text{mean}, s_{process}\rangle \cdot \text{cnorm}\left(\frac{x-UAL}{s_{gage}}\right) dx \ldots = 5.3344 \cdot 10^{-5}
$$

$$
+ \int_{min}^{LSL} \text{dnorm}\langle x, \text{mean}, s_{process}\rangle \cdot \text{cnorm}\left(\frac{LAL-x}{s_{gage}}\right) dx
$$

or, using the VBA function,

	Integration Limits		Acceptance		
	a	b	Limit		
Bad parts accepted above UAL	1020	1040	1017.94	4.99926E-06	
Bad parts accepted below LAL	960	980	−982.06	4.99926E-06	0.000010
Good parts rejected above UAL	980	1020	−1017.94	0.001019916	
Good parts rejected below LAL	980	1020	982.06	0.001019916	0.002040
Good parts accepted below UAL	980	1020	1017.94	0.998916742	
minus rejected below LAL	980	1020	982.06	0.001019916	0.997897
Bad parts rejected above UAL	1020	1040	−1017.94	2.6672E-05	
Bad parts rejected below LAL	960	980	982.06	2.6672E-05	0.000053
					1.000000

7

Multivariate Systems

Multivariate systems in which the process generates correlated quality characteristics are not this book's primary focus, but they arise sufficiently often to merit a brief overview and citations of potentially useful references.

Levinson and Cumbo (2000) describe a process that uses gaseous plasma to etch silicon wafers, and the depth of material removal is measured at nine positions on the wafer (Figure 7.1).

The measurements in the wafer's center and at its edges differ systematically but there is also correlation between them; more material removal at the center implies more etching at the edges as well. It is reasonable to expect the A positions to have the same mean, the B positions to share a different but correlated mean, and the C position to have a third mean. Any process of this nature suggests that a multivariate system will be present. Figure 7.2 shows a series of wafers in a tube furnace, in which reactive gas enters at one end and exits at the other. It is similarly reasonable to expect systematic differences between the wafers at the front, those in the middle, and those in the back.

Montgomery (1991, 322) cites bearings whose inner and outer diameters are correlated. This is a third example of a multivariate system, and this chapter will also discuss the issue of process performance indices for these systems.

Multivariate Normal Distribution

There are many similarities between the normal distribution and the multivariate normal distribution, although matrix algebra is necessary to handle calculations that involve the latter. If a process has p quality characteristics, an individual observation consists of a *vector* of p measurements. A sample of n is a $p \times n$ matrix, and the sample average is a vector of p elements as shown

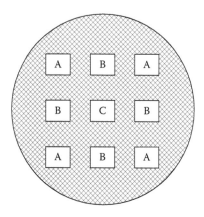

FIGURE 7.1
Measurement positions on silicon wafer.

in Equation (Set 7.1):

$$X = \begin{bmatrix} X_1 \\ X_2 \\ \vdots \\ X_p \end{bmatrix} \quad \text{Sample} \quad \begin{bmatrix} X_{11} & X_{12} & \cdots & X_{1n} \\ X_{21} & X_{22} & \cdots & X_{2n} \\ \vdots & \vdots & \vdots & \vdots \\ X_{p1} & X_{p2} & \cdots & X_{pn} \end{bmatrix} \quad \bar{X} = \begin{bmatrix} \dfrac{1}{n}\displaystyle\sum_{j=1}^{n} X_{1,j} \\ \dfrac{1}{n}\displaystyle\sum_{j=1}^{n} X_{2,j} \\ \vdots \\ \dfrac{1}{n}\displaystyle\sum_{j=1}^{n} X_{p,j} \end{bmatrix} \quad \text{(Set 7.1)}$$

The normal distribution has a mean μ and a variance σ^2, while the multivariate normal distribution has a *mean vector* μ and a *covariance matrix* Σ. For

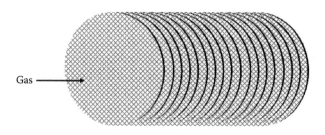

FIGURE 7.2
Another multivariate system: Silicon wafers in a tube furnace.

p quality characteristics,

$$
\mu = \begin{bmatrix} \text{Mean, quality characteristic 1} \\ \text{Mean, quality characteristic 2} \\ \cdots \\ \text{Mean, quality characteristic } p \end{bmatrix}
$$

$$
\Sigma = \begin{bmatrix} \sigma_{11} & \sigma_{12} & \cdots & \sigma_{1p} \\ \sigma_{21} & \sigma_{22} & \cdots & \sigma_{2p} \\ \vdots & \vdots & \vdots & \vdots \\ \sigma_{p1} & \sigma_{p2} & \cdots & \sigma_{pp} \end{bmatrix} \tag{Set 7.2}
$$

$$
\rho = \begin{bmatrix} 1 & \rho_{12} & \cdots & \rho_{1p} \\ \rho_{21} & 1 & \cdots & \rho_{2p} \\ \vdots & \vdots & \vdots & \vdots \\ \rho_{p1} & \rho_{p2} & \cdots & 1 \end{bmatrix}
$$

ρ_{ij} is the *correlation* between quality characteristics i and j. Only if it is zero are the two quality characteristics independent. Given these definitions, the multivariate probability density function is as shown in Equation (7.3):

$$
f(X) = \frac{1}{(2\pi)^{\frac{p}{2}} |\Sigma|^{\frac{1}{2}}} \exp\left(-\frac{1}{2}(X-\mu)^T \Sigma^{-1}(X-\mu)\right) \tag{7.3}
$$

where $|\Sigma|$ is the determinant of the covariance matrix and T is the transposition operator for matrices and vectors.[1] Note the similarity to the univariate normal probability density function and, if $p = 1$, Equation (7.3) becomes equal to it:

$$
f(x) = \frac{1}{\sqrt{2\pi}(\sigma^2)^{\frac{1}{2}}} \exp\left(-\frac{1}{2}(x-\mu)\sigma^{-2}(x-\mu)\right)
$$

The sample average in Equation (Set 7.1) is an estimate of the mean vector, and it is also necessary to estimate the covariance matrix.

Covariance and Correlation Matrices

As with the univariate normal distribution, standard practice is to use Greek letters for population parameters and their estimates, and English letters for sample statistics, for example,

$$S = \frac{1}{n-1}\sum_{j=1}^{n}(X_j - \bar{X})(X_j - \bar{X})^T \quad \text{sample of n}$$

(7.4a)

$$\hat{\Sigma} = \frac{1}{N-1}\sum_{j=1}^{N}(X_j - \bar{X})(X_j - \bar{X})^T \quad \text{population of N}$$

For computational purposes,

$$\hat{\Sigma}_{ik} = \frac{1}{N-1}\left[\sum_{j=1}^{N}x_{ij}x_{kj} - \frac{1}{N}\left(\sum_{j=1}^{N}x_{ij}\right)\left(\sum_{j=1}^{N}x_{kj}\right)\right]$$

(7.4b)

StatGraphics' documentation provides the alternative and equivalent formula (using x and y instead of x_i and x_k as shown here):

$$\hat{\Sigma}_{ik} = \frac{1}{N-1}\sum_{j=1}^{N}(x_{ij} - \bar{x}_i)(x_{kj} - \bar{x}_k)$$

(7.4c)

Holmes and Mergen (1998) offer an alternative method that relies on the mean square successive difference (MSSD), and a corresponding test for multivariate process stability:

$$\hat{\Sigma}_{ik,MSSD} = \frac{1}{2(N-1)}\sum_{j=2}^{N}(x_{i,j} - x_{i,j-1})(x_{k,j} - x_{k,j-1})$$

(7.4d)

The correlation matrix is then estimated as follows:

$$\hat{\rho} = \hat{D}^{-\frac{1}{2}}\hat{\Sigma}\hat{D}^{-\frac{1}{2}} \quad \text{where} \quad \hat{D} = \begin{bmatrix} \hat{\Sigma}_{11} & 0 & \cdots & 0 \\ 0 & \hat{\Sigma}_{22} & \cdots & 0 \\ \vdots & \vdots & \ddots & 0 \\ 0 & 0 & \cdots & \hat{\Sigma}_{pp} \end{bmatrix}$$

(7.5)

The Visual Basic for Applications (VBA) function Multivariate.bas reads an $n \times p$ array of data, calculates the mean vector, the covariance matrix per Equation (7.4c), and the mean square successive difference covariance matrix per Equation (7.4d). It writes them to the text file defined by OutPut_File, and the user can copy the results into Excel for further manipulation. It has been tested successfully by reproducing the covariance matrices in Table 7.1 (later in this chapter).

This section has so far defined the parameters of the multivariate normal distribution. The correlated nature of the quality characteristics means they cannot be treated as independent variables for SPC purposes, and

Montgomery (1991, 323–324) describes the problems that result from treating them as such. The next step is therefore to define the sample statistics for which control charts may be constructed. These charts test the null hypothesis that the process mean vector μ equals the nominal mean vector μ_0; their role is similar to that of the univariate x-bar and X charts.

d^2 and T^2 Charts

Recall that, for the univariate normal distribution, the standard normal deviate for a sample of n measurements with hypothetical mean μ_0 (e.g., the nominal for statistical process control [SPC] purposes) is

$$z = \frac{\bar{x} - \mu}{\frac{\sigma}{\sqrt{n}}} \Rightarrow z^2 = n(\bar{x} - \mu_0)\sigma^{-2}(\bar{x} - \mu_0)$$

The sum of the squares of p standard normal deviates follows a chi square distribution with p degrees of freedom, and the square of a single standard normal deviate as shown previously follows the chi square distribution with one degree of freedom. For the multivariate normal distribution for p quality characteristics,

$$d^2 = n(\bar{X} - \mu_0)^T \Sigma^{-1}(\bar{X} - \mu_0) \tag{7.6}$$

and d^2 follows the chi square distribution with p degrees of freedom.

In practice, it will be necessary to use the estimated covariance matrix and grand average vector (estimated mean vector) in place of μ_0 and Σ, an issue similar to that of empirical versus theoretical control charts. StatGraphics, in fact, uses different control limits for the two cases. For a theoretical control chart in which the mean vector and covariance matrix are given, the upper control limit is the $(1 - \alpha)$ quantile of the chi square distribution as described above. When the mean vector and covariance matrix are estimated from N data vectors,

$$UCL = \frac{p(N + 1)(N - 1)}{N(N - p)} F_{p;N-p} \tag{7.7}$$

where F is the upper $(1 - \alpha)$ quantile. When the process history is sufficiently extensive, the two cases become indistinguishable because $\lim_{N \to \infty} UCL = pF_{p;\infty} = \chi_p^2$ (Crofts, 1982). As an example, the tabulated 0.95 quantile of $F_{2;\infty}$ is 3.00, and two times this is 6.00. The tabulated 0.95 quantile of chi square with two degrees of freedom is 5.991.

The Hotelling T^2 chart is more widely known, and the calculation procedure is similar. The d^2 chart relies, however, on a covariance matrix that is estimated from the entire process history database. This approach is consistent with generally accepted practices for statistical process control, noting that univariate charts use a standard deviation whose basis is the entire process history. Suppose instead that an individual t test was performed for each sample, where s is the sample standard deviation:

$$t = \frac{\bar{x} - \mu_0}{\frac{s}{\sqrt{n}}} \Rightarrow t^2 = n(\bar{x} - \mu_0)s^{-2}(\bar{x} - \mu_0)$$

The resulting t statistics could then be plotted on a chart whose control limits were the $\alpha/2$ and $1 - \alpha/2$ quantiles of the t distribution with $n - 1$ degrees of freedom. The multivariate analogue to this procedure is

$$T^2 = n(\bar{X} - \mu_0)^T S^{-1}(\bar{X} - \mu_0) \sim \frac{(n-1)p}{(n-p)} F_{p;n-p} \tag{7.8}$$

and the upper control limit (or test statistic for a hypothesis test with a false alarm risk of α) is

$$\frac{(n-1)p}{(n-p)} F_{p;n-p;\alpha}$$

where the F statistic has an upper tail area of α, p degrees of freedom in the numerator (v_1) and $n - p$ degrees of freedom in the denominator (v_2). The key difference between d^2, which follows the chi square distribution, and T^2, which conforms to the F distribution as shown, is that Σ is the population covariance matrix while S is the covariance matrix of the sample.

Holmes and Mergen (1993) point out, in fact, that the covariance matrix for many "T^2" charts is usually that of the "base" data without subgrouping. This reference adds that this approach has the same drawbacks as use of the standard deviation of the entire database as opposed to the average of subgroup standard deviations to set the control limits of the univariate x-bar chart. This ties in with the issue of rational subgroups and process stability; if the process is stable, there should be no significant difference between

$$\hat{\Sigma} = \tfrac{1}{N-1}\sum_{j=1}^{N}(X_j - \bar{X})(X_j - \bar{X})^T \quad \text{and} \quad \hat{\Sigma}_{samples} = \frac{1}{k}\sum_{i=1}^{k} S_i.$$

The former would meanwhile form the basis of a process performance index while the latter would yield a process capability index. The reference offers the mean square successive difference method (Equation [7.4d]) as an alternative to either approach and concludes that, for a purely random process, $\hat{\Sigma}_{MSSD}$ will be very close to the estimate from the classical approach. The same authors (1998, 2004) then offer statistical tests for randomness and subgroup rationality.

Demonstration of these charts requires simulation of multivariate normal data, and the next section shows how to do this.

One hundred simulated pairs of measurements from the distribution

$$\Sigma = \begin{bmatrix} 3 & 1 \\ 1 & 2 \end{bmatrix} \text{ and } \mu = \begin{bmatrix} 3 \\ 1 \end{bmatrix}$$

are in the Multivariate_1 worksheet of simulate.xls. StatGraphics' multiple variable analysis finds an estimated mean vector

$$\hat{\mu} = \begin{bmatrix} 2.802 \\ 0.966 \end{bmatrix}$$

and estimated covariance matrix

$$\hat{\Sigma} = \begin{bmatrix} 3.140 & 1.109 \\ 1.109 & 1.950 \end{bmatrix}$$

MathCAD obtains the same results, and a quantile-quantile plot (Figure 7.3) of the d^2 values shows a moderately good fit to a chi square distribution with two degrees of freedom.

This is not an ideal quantile-quantile plot because the points do not scatter randomly around the best-fit line. This suggests that the simulation is not perfectly random, but it is good enough to illustrate the desired principles.[2] StatGraphics meanwhile fits the d^2 values to a chi square distribution with 1.93 degrees of freedom, noting that a chi square distribution with fractional degrees of freedom can be modeled by a gamma distribution. In a factory situation, the chi square test for goodness of fit also should be performed. The next step is to construct the control chart.

Multivariate Control Chart

The upper control limit for a d^2 chart with 2 degrees of freedom and a one-sided false alarm risk of 0.0027 is 11.83 per CHIINV(0.0027,2) in Microsoft Excel. StatGraphics generates the control chart in Figure 7.4 when given the estimated mean vector

$$\hat{\mu} = \begin{bmatrix} 2.802 \\ 0.966 \end{bmatrix}$$

and estimated covariance matrix

$$\hat{\Sigma} = \begin{bmatrix} 3.140 & 1.109 \\ 1.109 & 1.950 \end{bmatrix}$$

as standards. (StatGraphics calls it a T^2 chart, but it is technically a d^2 chart because a T^2 chart would use subgroup covariances as opposed to a pooled estimate.) Note that the point above the *UCL* is the same point that might be called an outlier in Figure 7.3.

$$d\_squared_i := \sum \left[\left(X^{<i>} - X\_bar \right) \right)^T \cdot S^{-1} \left(X^{<i>} - X\_bar \right) \right]$$

$$d\_ordered := sort(d\_squared) \qquad \qquad \chi_i := qchisq\left(\frac{i - 0.5}{n}, 2 \right)$$

$$\begin{bmatrix} b_0 \\ b_1 \\ r \end{bmatrix} := \begin{bmatrix} intercept(\chi, d\_ordered) \\ slope(\chi, d\_ordered) \\ corr(\chi, d\_ordered) \end{bmatrix} \qquad \begin{bmatrix} b_0 \\ b_1 \\ r \end{bmatrix} := \begin{bmatrix} -0.309 \\ 1.1485 \\ 0.9905 \end{bmatrix}$$

FIGURE 7.3
Quantile-quantile plot of d^2 versus chi square, 2 degrees of freedom.

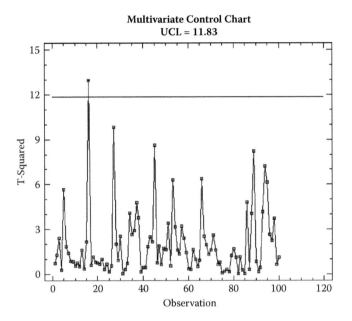

FIGURE 7.4
StatGraphics d^2 chart for control to standard.

As with the x-bar chart for univariate measurements, the d^2 chart will also accommodate results for samples. Consider samples of four while the standards for the mean vector and the covariance matrix are as shown above. Figure 7.5 shows the calculation for the first point on the control chart in Figure 7.6.

StatGraphics also offers a control ellipse (Figure 7.7) whose boundary consists of all points for which $d^2 = (X - \mu_0)^T \Sigma^{-1}(X - \mu_0) = 11.83$.

Sample   Population mean vector and covariance matrix

$$X = \begin{bmatrix} 2.71 & 2.822 & 0.246 & 2.461 \\ -0.118 & -0.425 & -0.642 & 1.436 \end{bmatrix} \quad \mu_0 := \begin{bmatrix} 2.802 \\ 0.966 \end{bmatrix} \Sigma_0 := \begin{bmatrix} 3.140 & 1.109 \\ 1.109 & 1.950 \end{bmatrix}$$

$n := 4$ Sample size $i := 1 .. 2$ Parameters

$$X\_bar_i := mean\left[\left(X^T\right)^{<i>}\right] \quad X\_bar = \begin{bmatrix} 2.06 \\ 0.063 \end{bmatrix} \quad \text{Sample average (mean vector)}$$

$$d\_squared := 4 \cdot \left(X\_bar - \mu_0\right)^T \cdot \Sigma_0^{-1} \cdot \left(X\_bar - \mu_0\right)$$

$$d\_squared := d\_squared_1 \quad \text{(convert vector to scalar)} \quad d\_squared = 1.757$$

FIGURE 7.5
Calculation of d^2 for a sample of 4.

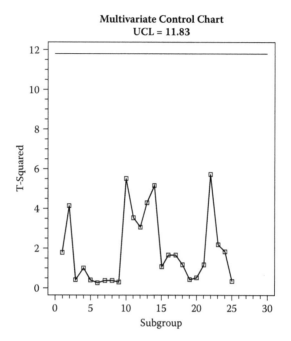

FIGURE 7.6
d^2 chart for samples of 4.

Consider the vector $[2.00, -3.574]^T$ as highlighted in Figure 7.7.

$$([2.00, -3.574] - [2.802, 0.966]) \begin{bmatrix} 3.140 & 1.109 \\ 1.109 & 1.950 \end{bmatrix}^{-1} \left(\begin{bmatrix} 2.00 \\ -3.574 \end{bmatrix} - \begin{bmatrix} 2.802 \\ 0.966 \end{bmatrix} \right)$$

$$= [-0.802, -4.540] \begin{bmatrix} 0.399 & -0.227 \\ -0.227 & 0.642 \end{bmatrix} \begin{bmatrix} -0.802 \\ -4.540 \end{bmatrix}$$

$$= [-0.802, -4.540] \begin{bmatrix} 0.7106 \\ -2.733 \end{bmatrix} = 11.84$$

MathCAD obtains 11.83 when all the significant figures in the mean vector and covariance matrix are present. An important lesson of the control ellipse is that it shows the correlation, or interdependence, between the two measurements. The outcome $x_2 = 4$ is, for example, a reasonable outcome when $x_1 = 5$ but not when $x_1 = 0$. If the two quality characteristics were independent, this relationship would not exist, so they could indeed be plotted independently.

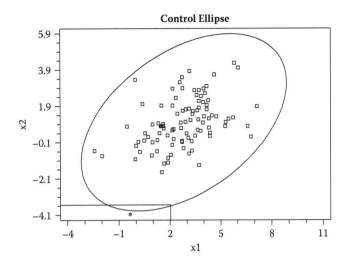

FIGURE 7.7
Control ellipse, bivariate normal data.

Sauers (1997) shows how to set up a Hotelling T^2 chart in Corel Quattro Pro. We have not attempted to do this, but relevant Excel functions include MMULT (matrix multiplication), TRANSPOSE (matrix transposition), MINVERSE (matrix inversion), and MDETERM (matrix determinant). Mason, Chou, and Young (2001) meanwhile take up the issue of multivariate batch processes.

It is, however, possible to set up a d^2 chart on a spreadsheet without using any matrix operations whatsoever, and this is the subject of the next section.

Deployment to a Spreadsheet: Principal Component Method

The previous section has shown that calculation of d^2 or T^2 is very computationally intensive. It requires algorithms for matrix inversion or matrix determinants and, although MathCAD can perform these functions, it is not particularly suitable for deployment to the shop floor. Neither are StatGraphics and Minitab even though both can perform the "vernier" part of the job in the statistical practitioner's office. This section will address the "stone ax" part of the job, or something simple that can in fact be deployed on a spreadsheet if suitable commercial software is not available.[3]

Principal Components

Johnson and Wichern (1992, Chapter 8) describe principal components as follows. Given p quality characteristics and a covariance matrix with eigenvalues $\lambda_1 \geq \lambda_2 \geq \cdots \lambda_p$ and corresponding eigenvectors e_k ($k = 1$ through p):

1. Principal component $Y_k = e_k^T X = \sum_{i=1}^p e_{ik} X_i$.
2. $Var(Y_k) = e_k^T \Sigma e_k = \lambda_k$; that is, the variance of the kth principal component is its eigenvalue.
3. $Cov(Y_i, Y_k) = 0$ where $i \neq k$, that is, the principal components are independent and orthogonal. If the underlying distribution is multivariate normal, the principal components will behave like independently distributed normal variables.
4. Note also that the eigenvectors are standardized such that $e_k^T e_k = 1$.

In the case of the 100 simulated data, StatGraphics and Minitab obtain the following sets of coefficients (non-standardized)[4] for the linear combinations:

$$\begin{bmatrix} 0.8581 \\ 0.5135 \end{bmatrix}$$

and

$$\begin{bmatrix} -0.5135 \\ 0.8581 \end{bmatrix}$$

This meets the requirement $c^T c = 1$, for example,

$$[0.8581, 0.5135] \begin{bmatrix} 0.8581 \\ 0.5135 \end{bmatrix} = 0.7363 + 0.2637 = 1.$$

Furthermore, the respective eigenvalues are 3.804 and 1.287, and the kth principal component accounts for fraction

$$\frac{\lambda_k}{\sum_{i=1}^p \lambda_i}$$

of the total variation. In this case, the first principal component accounts for

$$\frac{3.804}{3.804 + 1.287} = 74.7\%$$

of the variation.

Recalling that

$$\hat{\mu} = \begin{bmatrix} 2.802 \\ 0.966 \end{bmatrix}$$

and

$$\hat{\Sigma} = \begin{bmatrix} 3.140 & 1.109 \\ 1.109 & 1.950 \end{bmatrix},$$

the first principal component should have

$$\hat{\mu} = [0.8581,\ 0.5135] \begin{bmatrix} 2.802 \\ 0.966 \end{bmatrix} = 2.900$$

and

$$\sigma^2 = [0.8581,\ 0.5135] \begin{bmatrix} 3.140 & 1.109 \\ 1.109 & 1.950 \end{bmatrix} \begin{bmatrix} 0.8581 \\ 0.5135 \end{bmatrix} = [0.8581,\ 0.5135] \begin{bmatrix} 3.264 \\ 1.953 \end{bmatrix} = 3.804$$

or $\sigma = 1.950$. This is what StatGraphics obtains for the estimated mean and standard deviation of the first principal component and the quantile-quantile plot shows a good fit to a normal distribution. For the second principal component,

$$\hat{\mu} = [-0.5135,\ 0.8581] \begin{bmatrix} 2.802 \\ 0.966 \end{bmatrix} = -0.6099 \text{ and}$$

$$\sigma^2 = [-0.5135,\ 0.8581] \begin{bmatrix} 3.140 & 1.109 \\ 1.109 & 1.950 \end{bmatrix} \begin{bmatrix} -0.5135 \\ 0.8581 \end{bmatrix}$$

$$= [-0.5135,\ 0.8581] \begin{bmatrix} -0.6608 \\ 1.1038 \end{bmatrix} = 1.287$$

or $\sigma = 1.134$, and StatGraphics obtains -0.6101 and 1.134, respectively.

This section has described so far how the principal components follow independent normal distributions whose means and standard distributions can be calculated through the indicated matrix operations. The next step is to use the *standardized* principal components, which are similar in concept to standard normal deviates, to deploy the d^2 chart without recourse to matrix operations. The mean-centered principal components also work for this purpose.

d^2 Chart from Standardized or Mean-Centered Principal Components

Per Mason and Young (2005), where Y_k is the kth component,

$$d^2 = \sum_{k=1}^{p} \frac{Y_k^2}{\lambda_k} \tag{7.9}$$

As described by Johnson and Wichern (1992, 361–262), the ellipsoid

$$c^2 = (x-\mu)^T \Sigma^{-1}(x-\mu) = \sum_{k=1}^{p} \frac{Y_k^2}{\lambda_k}$$

where $Y_k = e_k^T x$. These statements require, however, that Y_k be either a mean-centered principal component or a standardized principal component, which is by definition mean-centered.

When the Standardize option is selected in StatGraphics and Correlation is selected in Minitab, the programs return 1.4482 and 0.5518 as the eigenvalues for the standardized principal components

$$Y_1 = \frac{\sqrt{2}}{2} Z_1 + \frac{\sqrt{2}}{2} Z_2$$

and

$$Y_2 = \frac{\sqrt{2}}{2} Z_1 - \frac{\sqrt{2}}{2} Z_2 \quad \text{where } Z_k = \frac{X_k - \mu_k}{\sqrt{\sigma_{kk}}}$$

Recalling that

$$\hat{\mu} = \begin{bmatrix} 2.802 \\ 0.966 \end{bmatrix}$$

and

$$\hat{\Sigma} = \begin{bmatrix} 3.140 & 1.109 \\ 1.109 & 1.950 \end{bmatrix}, \ Z_1 = \frac{X_1 - 2.802}{\sqrt{3.140}} \text{ and } Z_2 = \frac{X_2 - 0.966}{\sqrt{1.950}}$$

StatGraphics returns −0.5854 and 0.5121 for the standardized principal components of the first measurement vector.

$$d^2 = \frac{(-0.5854)^2}{1.4482} + \frac{0.5121^2}{0.5518} = 0.712,$$

which matches the previous result from matrix operations.

Mean-centered but nonstandardized principal components also work, and are computationally even simpler. The nonstandardized principal components of the first measurement are 2.265 and −1.493, and the eigenvalues of the covariance matrix are 3.804 and 1.287. Recall also that the means of the principal components are 2.900 and 0.6099 (0.610), respectively.

$$d^2 = \frac{(2.265 - 2.900)^2}{3.804} + \frac{(-1.493 - -0.610)}{1.287} = 0.712 \text{ (matches previous results)}$$

This is easily deployed on a spreadsheet as

$$d^2 = \sum_{k=1}^{p} \frac{(Y_k - \mu_k)^2}{\lambda_k}.$$

In summary, while relatively sophisticated software is required for the "vernier" part of the job—specifically, calculation of the eigenvalues, eigenvectors, and coefficients for the principal components—a spreadsheet can perform the "stone ax" role by allowing the operator to enter the raw measurements and get the multivariate test statistic without recourse to any matrix operations whatsoever. Conditional formatting of the cell can turn it red if d^2 exceeds the control limit.

When an out-of-control condition is detected, the next step is to determine which quality characteristic or group of characteristics is responsible. StatGraphics offers T^2 decomposition, which measures each variable's contribution to the overall test statistic. Mason Tracy, and Young (1995, 1997) also describe approaches for interpretation of the out-of-control signal.

Multivariate Process Performance Index

Recall that the process performance index is a statement to the customer about the anticipated nonconforming fraction from the process. In the case of a multivariate system, all the correlated quality characteristics must be within their specification limits. For the bivariate system in which x and y are the quality characteristics and $f(x,y)$ is the bivariate probability density function per Equation (7.3), the estimated conforming fraction is

$$P = \int_{LCL,y}^{UCL,y} \int_{LCI,x}^{UCL,x} f(x,y)\,dx\,dy \text{ and then } P_p = \frac{1}{3}\Phi^{-1}(P) \qquad (7.10)$$

The Automotive Industry Action Group (2005, 144–145) provides a reference for this approach, which would presumably extend to triple integration of trivariate distributions and so on.

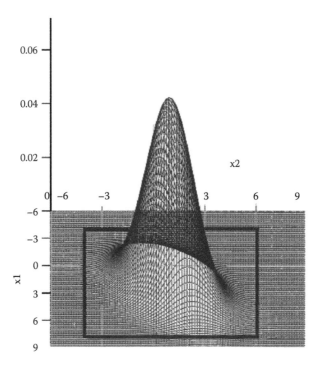

FIGURE 7.8A
Process performance index, bivariate normal process.

Consider the example in which

$$\hat{\mu} = \begin{bmatrix} 2.802 \\ 0.966 \end{bmatrix}$$

and

$$\hat{\Sigma} = \begin{bmatrix} 3.140 & 1.109 \\ 1.109 & 1.950 \end{bmatrix}$$

and assume that the specification limits are [−4,8] and [−4,6] for x_1 and x_2, respectively. As shown by Figure 7.8A (MathCAD three-dimensional plot), the integrated volume within these limits is in specification, and the volume outside it is nonconforming. The equivalent process performance index is 0.956.

Figure 7.8B shows how StatGraphics displays the capability analysis. A nonconforming fraction of 0.00208 (2080 defects per million opportunities [DPMO]) is roughly consistent with a conforming fraction of 0.99793. The key point is that both quality characteristics must be within their respective specifications, and the joint probability that this will happen is defined by the multivariate probability density function.

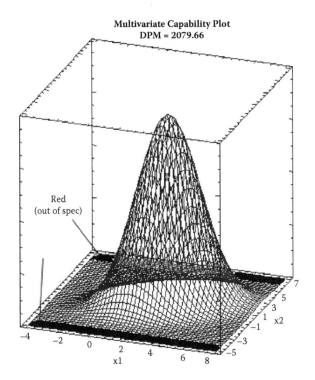

FIGURE 7.8B
Bivariate process performance index, StatGraphics.

Integration may admittedly become computationally impractical for large numbers of quality characteristics. Holmes and Mergen (1998a and 2001) offer an alternative approach that might be useful, and Wang and Chen (1998) describe a method that uses the geometric average of the univariate indices for the principal components.

This chapter has described the multivariate analogues of the X and x-bar charts for the process mean, and the process performance index. There are also multivariate analogues of the R and s charts for process variation, and their function is to detect changes in the covariance matrix.

Control Charts for the Covariance Matrix

A change in process variance is usually an undesirable increase that makes it more likely that work will be out of specification. The effects of a change in the covariance matrix are not as straightforward, although increases in the diagonal elements correspond to greater variation and therefore poorer quality. Changes in off-diagonal elements reflect changes in the correlation between different quality characteristics. Several charts for the covariance matrix are available, and Vargas and Lagos (2007) compare four of them: the generalized variance chart, W or S^2 chart, G chart, and RG (Robust G) chart.

Generalized Variance Chart

StatGraphics uses the *generalized variance chart*, which plots the determinant of the sample covariance matrix as described by Montgomery (1991, 330–331).[5] The control limits rely on a normal approximation to the distribution of $|S|$ as shown in Equation (Set 7.11). For p parameters and a sample of n,

Expected determinant and chart center line: $E(|S|) = b_1 |\Sigma|$

Variance of determinant: $V(|S|) = b_2 |\Sigma|^2 \Rightarrow \sigma_{|S|} = \sqrt{b_2} |\Sigma|$

Control limits: $b_1 |\Sigma| \pm 3\sqrt{b_2} |\Sigma| = |\Sigma|\left(b_1 \pm 3\sqrt{b_2}\right)$

$$\text{(Set 7.11)}$$

$$b_1 = \frac{1}{(n-1)^p} \prod_{i=1}^{p} (n-i)$$

$$b_2 = \frac{1}{(n-1)^{2p}} \prod_{i=1}^{p} (n-i) \left[\prod_{j=1}^{p} (n-j+2) - \prod_{j=1}^{p} (n-j) \right]$$

When Σ is estimated from the process history, as is often the case, the reference recommends replacement of $|\Sigma|$ with

$$\frac{|\hat{\Sigma}|}{b_1}$$

as an unbiased estimator of $|\Sigma|$. In this case, the control limits become

$$|\hat{\Sigma}|\left(1 \pm 3\frac{\sqrt{b_2}}{b_1}\right).$$

W Chart or S^2 Chart

For p quality characteristics and a sample size of n, compute

$$W_i = -p(n-1) - (n-1)\ln|\Sigma_0| + (n-1)tr\left(\frac{S_i}{\Sigma_0}\right) \tag{7.12}$$

where tr is the trace of the matrix, the sum of the diagonal elements. The W statistic follows the chi square distribution with $p(p + 1)/2$ degrees of freedom. Montgomery (1991, 330–331) adds that, when the population covariance must be estimated from the process history,

$$\frac{|\hat{\Sigma}|}{b_1}$$

is an unbiased estimator of $|\Sigma_0|$ as shown previously.

G Chart

Levinson, Holmes, and Mergen (2002) introduce a chart whose basis is a test for equality of two covariance matrices per Kramer and Jensen (1969). The concept is similar to the test for equality of standard deviations from two samples in the context of the t test. Given p quality characteristics, samples of sizes n_1 and n_2, sample covariance matrices S_1 and S_2, and pooled covariance matrix

$$S = \frac{(n_1 - 1)S_1 + (n_2 - 1)S_2}{n_1 + n_2 - 2}$$

(note the similarity of the last to the pooled variance estimate for univariate t tests), the null hypothesis $\Sigma_1 = \Sigma_2$ is tested as shown in Equation (Set 7.13).

$$M = (n_1 + n_2 - 2)\ln |S| - (n_1 - 1)\ln |S_1| - (n_2 - 1)\ln |S_2| = \ln\left(\frac{|S|^{n_1+n_2-2}}{|S_1|^{n_1-1}|S_2|^{n_2-1}}\right)$$

$$m = 1 - \left(\frac{1}{n_1 - 1} + \frac{1}{n_2 - 1} - \frac{1}{n_1 + n_2 - 2}\right) \times \left(\frac{2p^2 + 3p - 1}{6(p + 1)}\right) \qquad \text{(Set 7.13)}$$

$$G = M \times m$$

If G exceeds the $1 - \alpha$ quantile of the chi square statistic with $p(p+1)/2$ degrees of freedom, reject the null hypothesis that $\Sigma_1 = \Sigma_2$ at the α significance level. Holmes and Mergen (1998) convert this into a test for process stability by letting $S_{ik} = \frac{1}{N-1}[\sum_{j=1}^{N} x_{ij}x_{kj} - \frac{1}{N}(\sum_{j=1}^{N} x_{ij})(\sum_{j=1}^{N} x_{kj})]$ as described in Equation (7.4b) and $S_{ik,MSSD} = \frac{1}{2(N-1)}\sum_{j=2}^{N}(x_{i,j} - x_{i,j-1})(x_{k,j} - x_{k,j-1})$ per Equation (7.4d). The G test is then used to test the hypothesis that the process is stable.

The same reference presents a table of percentage by weight of five screens (A, B, C, D, E) for determination of particle size (see Table 7.1A), with a total of 35 measurements ($p = 5$, $n = 35$). S from Equation (7.4b) is, per MathCAD, StatGraphics (Describe/Multivariate Methods/Multiple Variable Analysis), and Minitab 16 (Stat/Basic Statistics/Correlation) as shown in Table 7.1A.

TABLE 7.1A

Covariance Matrix, Particle Screening

1.5379	0.9087	1.2076	1.1318	0.3358
0.9087	6.8634	0.3894	2.3727	−4.0653
1.2076	0.3894	4.9601	3.8241	1.7776
1.1318	2.3727	3.8241	5.4678	−0.0121
0.3358	−4.0653	1.7776	−0.0121	5.4668

For S_{MSSD} per MathCAD, see Table 7.1B.

In this case, the pooled covariance matrix (see Table 7.1C) is

$$S_{pooled} = \frac{(n-1)S_1 + (n-1)S_2}{2n-2} = \frac{S + S_{MSSD}}{2}$$

noting that $n_1 = n_2 = n$.

Figure 7.9A shows the calculation of G, and the fact that it exceeds the 0.995 quantile of the chi square statistic with $5(5 + 1)/2 = 15$ degrees of freedom shows that the process is not stable. The quantile-quantile plot of d^2 against χ^2 with 15 degrees of freedom (Figure 7.9B) also shows that there is a serious problem with the underlying assumptions.

Holmes and Mergen (2004) extend this approach to test for the rationality of multivariate subgroups, and Levinson, Holmes, and Mergen (2002) use it to create a control chart for subgroup covariances as follows.

Given k samples of size n_2, let $S_{2,i}$ be

$$\frac{1}{n_2 - 1} \sum_{j=1}^{n} (X_j - \bar{X})(X_j - \bar{X})^T$$

where $\bar{X}$ is the mean vector of the ith sample. Then

$$S_1 = \frac{1}{k} \sum_{i=1}^{k} S_{2,i},$$

or simply the average of the sample covariance matrices. Then apply Equation (Set 7.14) (G chart for sample covariances) to each subgroup, noting that S_1

TABLE 7.1B

Covariance Matrix, Particle Screening, MSSD

1.5157	0.2919	1.4714	1.3284	0.9437
0.2919	1.1442	0.6118	1.1319	0.2802
1.4714	0.6118	3.8272	3.2405	1.0776
1.3284	1.1319	3.2405	4.8523	0.6379
0.9437	0.2802	1.0776	0.6379	2.2276

TABLE 7.1C

Pooled Covariance Matrix, Particle Screening

1.5268	0.6003	1.3395	1.2301	0.6397
0.6003	4.0038	0.5006	1.7523	-1.8926
1.3395	0.5006	4.3936	3.5323	1.4276
1.2301	1.7523	3.5323	5.16	0.3129
0.6397	-1.8926	1.4276	0.3129	3.8472

has $k(n_2 - 1)$ degrees of freedom while each subgroup $S_{2,i}$ has $n_2 - 1$ degrees of freedom.

$$S_{pooled} = \frac{k(n_2 - 1)S_1 + (n_2 - 1)S_2}{k(n_2 - 1) + (n_2 - 1)} = \frac{kS_1 + S_2}{k+1}$$

$$M = (k+1)(n_2 - 1)\ln|S_{pooled}| - k(n_2 - 1)\ln|S_1| - (n_2 - 1)\ln|S_2|$$

$$= (n_2 - 1)\big((k+1)\ln|S_{pooled}| - k\ln|S_1| - \ln|S_2|\big)$$

$$m = 1 - \left(\frac{1}{k(n_2-1)} + \frac{1}{(n_2-1)} - \frac{1}{(k+1)(n_2-1)}\right)\left(\frac{2p^2+3p-1}{6(p+1)}\right) \quad \text{(Set 7.14)}$$

$$= 1 - \frac{1}{n_2-1}\left(\frac{1}{k} + 1 - \frac{1}{k+1}\right)\left(\frac{2p^2+3p-1}{6(p+1)}\right)$$

$$G = M \times m$$

Then set the control limits equal to the $\alpha/2$ and $1 - \alpha/2$ quantiles of the chi square distribution with $p(p + 1)/2$ degrees of freedom.

$$M := (2 \cdot n - 2) \cdot \ln(|S_{pooled}|) - (n-1) \cdot \ln(|S|) - (n-1) \cdot \ln(|S_{MSSD}|)$$

$$M = 41.507$$

$$m := 1 - \left(\frac{2}{n-1} - \frac{1}{2 \cdot n - 2}\right)\left[\frac{2 \cdot p^2 + 3 \cdot p - 1}{6 \cdot (p+1)}\right] \quad m = 0.922$$

$$G := m \cdot M \qquad G = 38.251 \qquad \text{qchisq}(.995,15) = 32.801$$

(The reference gets $M = 17.949 \times \ln(10) = 41.329$, and $G = 38.064$.)

FIGURE 7.9A

G test for process stability, particle screening.

$$\text{d\_squared}_j := \sum \left[\left(\left(X^{\langle j \rangle} - X\_bar \right) \right)^T \cdot S^{-1} \left(X^{\langle j \rangle} - X\_bar \right) \right]$$

$$\text{d\_ordered} := \text{sort(d\_squared)} \qquad\qquad X_j := \text{qchisq}\left(\frac{j - 0.5}{n}, 5 \right)$$

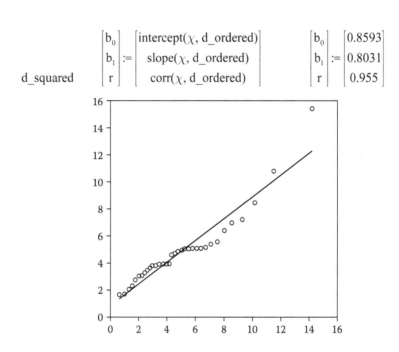

$$\text{d\_squared} \qquad \begin{bmatrix} b_0 \\ b_1 \\ r \end{bmatrix} := \begin{bmatrix} \text{intercept}(\chi, \text{d\_ordered}) \\ \text{slope}(\chi, \text{d\_ordered}) \\ \text{corr}(\chi, \text{d\_ordered}) \end{bmatrix} \qquad \begin{bmatrix} b_0 \\ b_1 \\ r \end{bmatrix} := \begin{bmatrix} 0.8593 \\ 0.8031 \\ 0.955 \end{bmatrix}$$

FIGURE 7.9B
d^2 quantile-quantile plot, particle screening.

Consider, for example, the 100 simulated multivariate data (Multivariate_1 in the simulated data spreadsheet). The first sample of 4 is

2.71	−0.118
2.822	−0.425
0.246	−0.642
2.461	1.436

and its covariance matrix is

1.4848	0.4466
0.4466	0.8844

Figure 7.10 shows the calculation for the first covariance matrix in MathCAD, followed by calculation of the G statistics.

$$\begin{bmatrix} N \\ n_2 \\ k \\ p \end{bmatrix} = \begin{bmatrix} 100 \\ 4 \\ 25 \\ 2 \end{bmatrix}$$

$$S_2(i) := \frac{1}{n_2 - 1} \sum_{j=n_2 \cdot (i-1)+1}^{n_2 \cdot i} \left[x^{<j>} - \frac{1}{n_2} \cdot \sum_{1=n_2 \cdot (i-1)+1}^{n_2 \cdot i} x^{<1>} \right]$$

(all one line in the actual MathCAD sheet)

$$X \left[x^{<j>} - \frac{1}{n_2} \cdot \sum_{1=n_2 \cdot (i-1)+1}^{n_2 \cdot i} x^{<1>} \right]^T$$

$$S_2(1) = \begin{bmatrix} 1.4848 & 0.4466 \\ 0.4466 & 0.8844 \end{bmatrix} \quad S_1 := \frac{1}{k} \cdot \sum_{j=1}^{k} S_2(j) \quad S_1 = \begin{bmatrix} 3.5159 & 1.3057 \\ 1.3057 & 1.8528 \end{bmatrix}$$

$$S_{pooled}(i) := \frac{k \cdot S_1 + S_2(i)}{k+1} \quad m := 1 - \frac{1}{n_2 - 1} \cdot \left(\frac{1}{k} + 1 - \frac{1}{k+1} \right) \cdot \left[\frac{2p^2 + 3 \cdot p - 1}{6 \cdot (p+1)} \right] \quad m = 0.759$$

$$M(i) := (n_2 - 1) \cdot [(k+1) \cdot \ln(|S_{pooled}(i)|) - k \cdot \ln(|S_1|) - \ln(|S_2(i)|)]$$

$$i := 1..k \qquad\qquad G_i := m \cdot M(i)$$

FIGURE 7.10
Calculation of G statistics.

The G statistics should conform to a chi square distribution with $2(2+1)/2 = 3$ degrees of freedom, and Figure 7.11 shows that they do so at least roughly. For a control chart with a 2-sided false alarm risk of 1%, the control limits would be the 0.005 and 0.995 quantiles of chi square with 3 degrees of freedom, or 0.072 and 12.838, respectively.

The RG Chart

Vargas and Lagos (2007) offer an improvement on the G chart by using a robust estimator of the covariance matrix as opposed to the average

$$S_1 = \frac{1}{k} \sum_{i=1}^{k} S_{2,i}.$$

This is specifically the minimum volume ellipsoid (MVE) estimator whose basis is the entire dataset, which is in turn calculated as shown in Vargas (2003). The authors add that the generalized variance and RG charts are good at detecting increases in the standard deviations of the quality characteristics, while the S^2, G, and RG charts were effective for detecting large process

$$G_{ordered} := \text{sort}(G) \qquad axis_i := \text{qchisq}\left(\frac{i-0.5}{k}, 3\right)$$

$$\begin{bmatrix} b_0 \\ b_1 \\ r \end{bmatrix} := \begin{bmatrix} \text{intercept}(axis, G_{ordered}) \\ \text{slope}(axis, G_{ordered}) \\ \text{corr}(axis, G_{ordered}) \end{bmatrix} \qquad \begin{bmatrix} b_0 \\ b_1 \\ r \end{bmatrix} := \begin{bmatrix} 0.555 \\ 0.963 \\ 0.971 \end{bmatrix}$$

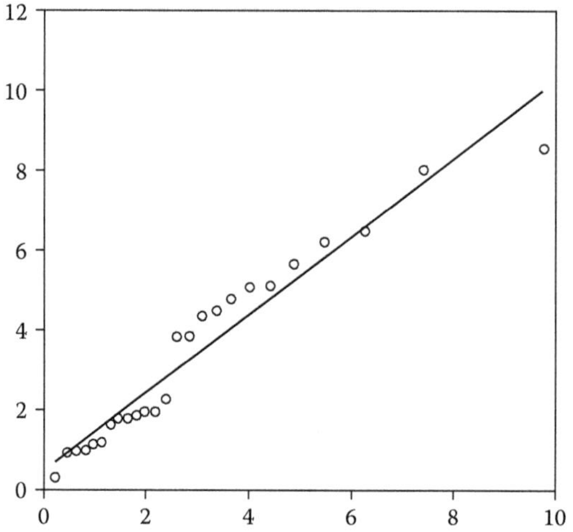

FIGURE 7.11
Quantile-quantile plot, ordered Gs versus χ^2.

improvements in simulated datasets. We have not had the opportunity to try this algorithm, but we include these references for informational purposes.

Endnotes

1. Transposition: $y_{ij} = x_{ji}$.
2. A five-parameter system (Multivariate_2 in the simulated data spreadsheet) meanwhile yields a more "textbook" Q-Q plot of d^2 versus chi square with 5 degrees of freedom.
3. Custom QC from Stochos.com accepts data in a spreadsheet format and can prepare a T^2 chart. A Google search on multivariate SPC also leads to BaRan Systems, which offers an Excel add-in (http://baran-systems.com/Products/Advanced%20SQC%20for%20Excel/index_concept.htm) that handles multivariate applications. See also QualStat at http://www.qualstat.com/. We have not evaluated any of these programs, and these sources are provided as information only.

4. In StatGraphics, make sure the Standardize box is not checked in the analysis options. In Minitab, select the Covariance Matrix box as opposed to the Correlation Matrix option.
5. Montgomery (1991) and Vargas and Lagos (2007) cite F. B. Alt, 1985. "Multivariate Quality Control" in *Encyclopedia of Statistical Sciences*, volume 6, edited by N. L. Johnson and S. Kotz (New York: John Wiley, 1985).

Glossary

d^2: Test statistic for the mean vector in multivariate normal control charts

e: Eigenvector, or matrix of eigenvectors

$f(x)$: Probability density function

$F(y)$: Cumulative density function (integral of $f(x)$ from its lower limit to y)

GRR: Gage reproducibility and repeatability

HDS: Historical dataset

I: Identity matrix with diagonal elements 1 and off-diagonal elements 0

k: Tolerance factor for population fraction P

LAL: Lower acceptance limit

LCL: Lower control limit

LSL: Lower specification limit

$N(\mu,\sigma^2)$: Normal distribution with mean μ and variance σ^2

p: Nonconforming fraction

pdf: Probability density function

PTCC: Percent of tolerance consumed by gage capability, = P/T ratio

$S(x)$: Survivor function in reliability statistics, $=1 - F(x)$

SPC: Statistical process control

T: Transposition operator for matrix operators

UAL: Upper acceptance limit

UCL: Upper control limit

USL: Upper specification limit

VBA: Visual Basic for Applications

z: Standard normal deviate $(x - \mu)/\sigma$

α: alpha risk, Type I risk, or producer's risk. "The boy who cried wolf."

α: shape parameter for the gamma distribution

β: beta risk, Type II risk, or consumer's risk. "The boy didn't see the wolf."

β: shape parameter for the Weibull distribution

$\Gamma(x)$: Gamma function; for integers, $\Gamma(x) = (x - 1)!$

γ: gamma, the power of a test. $\gamma = 1 - \beta$ where β is the Type II risk.

γ: gamma, the confidence level for a tolerance interval

γ: gamma, the scale parameter of the gamma distribution

δ: threshold parameter

θ: scale parameter of the Weibull distribution

θ: mean of an exponential distribution (mean time between failures)

Λ: Log likelihood ratio statistic

λ: Eigenvalue, or vector of eigenvalues

λ: Hazard rate of an exponential distribution

μ: Population mean or (multivariate statistics), mean vector

ν: Degrees of freedom

ρ: Correlation matrix, multivariate normal distribution

Σ: Covariance matrix, multivariate normal distribution

σ: Population standard deviation (σ^2 = population variance)

Φ(z): Cumulative standard normal density function

Φ^{-1} (p): Inverse cumulative standard normal density function for proportion p

Appendix A: Control Chart Factors

Formulas for Control Chart Factors

Reference: ASTM (1990, 91–94)

Factors for central line

s chart $c_4 = \sqrt{\dfrac{2}{n-1}}\dfrac{\Gamma\left(\frac{n}{2}\right)}{\Gamma\left(\frac{n-1}{2}\right)}$ $\hat{\sigma} = \dfrac{\bar{s}}{c_4}$

R chart $\hat{\sigma} = \dfrac{\bar{R}}{d_2}$

$d_2 = \displaystyle\int_{-\infty}^{\infty}\left(1-\Phi(x_1)^n - (1-\Phi(x_1))^n\right)dx$ $\Phi(x) = $ cumulative standard normal distribution

$d_3 = \sqrt{2\displaystyle\int_{-\infty}^{\infty}\int_{-\infty}^{x_1}\left[1-\Phi(x_1)^n - \left(1-\Phi(x_n)\right)^n + \left(\Phi(x_1)-\Phi(x_n)^n\right)\right]dx_n\,dx_1 - d_2^2}$

Factors for the s chart

Theoretical: $\begin{aligned} B_5 &= c_4 - 3\sqrt{1-c_4^2} \\ B_6 &= c_4 + 3\sqrt{1-c_4^2} \end{aligned}$

Empirical: $B_3 = \dfrac{B_5}{c_4}$ $B_4 = \dfrac{B_6}{c_4}$

noting that $\hat{\sigma} = \dfrac{\bar{s}}{c_4} \Rightarrow B_3\bar{s} = \dfrac{B_5}{c_4}\bar{s} = B_5\hat{\sigma}$

Factors for the R chart

Theoretical: $\begin{aligned} D_1 &= d_2 - 3d_3 \\ D_2 &= d_2 + 3d_3 \end{aligned}$

Empirical: $D_3 = \dfrac{D_1}{d_2}$ $D_4 = \dfrac{D_2}{d_2}$

noting that $\hat{\sigma} = \dfrac{\bar{R}}{d_2} \Rightarrow D_3\bar{R} = \dfrac{D_1}{d_2}\bar{R} = D_1\hat{\sigma}$

Factors for the x-bar chart

$$A = \dfrac{3}{\sqrt{n}} \quad \mu \pm \dfrac{3\sigma}{\sqrt{n}} = \mu \pm A\sigma$$

$$A_2 = \dfrac{3}{d_2\sqrt{n}} \quad \hat{\sigma} = \dfrac{\bar{R}}{d_2} \Rightarrow \bar{\bar{x}} \pm \dfrac{3\hat{\sigma}}{\sqrt{n}} = \bar{\bar{x}} \pm A_2\bar{R}$$

$$A_3 = \dfrac{3}{c_4\sqrt{n}} \quad \hat{\sigma} = \dfrac{\bar{s}}{c_4} \Rightarrow \bar{\bar{x}} \pm \dfrac{3\hat{\sigma}}{\sqrt{n}} = \bar{\bar{x}} \pm A_3\bar{s}$$

n	A	A2	A3	c4	1/c4	B3	B4	B5	B6	d2	d3	1/d2	D1	D2	D3	D4
2	2.121	1.880	2.659	0.7979	1.2533	0.000	3.267	0.000	2.606	1.128	0.853	0.8862	0.000	3.686	0.000	3.267
3	1.732	1.023	1.954	0.8862	1.1284	0.000	2.568	0.000	2.276	1.693	0.888	0.5908	0.000	4.358	0.000	2.575
4	1.500	0.729	1.628	0.9213	1.0854	0.000	2.266	0.000	2.088	2.059	0.880	0.4857	0.000	4.698	0.000	2.282
5	1.342	0.577	1.427	0.9400	1.0638	0.000	2.089	0.000	1.964	2.326	0.864	0.4299	0.000	4.918	0.000	2.114
6	1.225	0.483	1.287	0.9515	1.0509	0.030	1.970	0.029	1.874	2.534	0.848	0.3946	0.000	5.079	0.000	2.004
7	1.134	0.419	1.182	0.9594	1.0424	0.118	1.882	0.113	1.806	2.704	0.833	0.3698	0.205	5.204	0.076	1.924
8	1.061	0.373	1.099	0.9650	1.0362	0.185	1.815	0.179	1.751	2.847	0.820	0.3512	0.388	5.307	0.136	1.864
9	1.000	0.337	1.032	0.9693	1.0317	0.239	1.761	0.232	1.707	2.970	0.808	0.3367	0.547	5.394	0.184	1.816
10	0.949	0.308	0.975	0.9727	1.0281	0.284	1.716	0.276	1.669	3.078	0.797	0.3249	0.686	5.469	0.223	1.777
11	0.905	0.285	0.927	0.9754	1.0253	0.321	1.679	0.313	1.637	3.173	0.787	0.3152	0.811	5.535	0.256	1.744
12	0.866	0.266	0.886	0.9776	1.0230	0.354	1.646	0.346	1.610	3.258	0.778	0.3069	0.923	5.594	0.283	1.717
13	0.832	0.249	0.850	0.9794	1.0210	0.382	1.618	0.374	1.585	3.336	0.770	0.2998	1.025	5.647	0.307	1.693
14	0.802	0.235	0.817	0.9810	1.0194	0.406	1.594	0.399	1.563	3.407	0.763	0.2935	1.118	5.696	0.328	1.672
15	0.775	0.223	0.789	0.9823	1.0180	0.428	1.572	0.421	1.544	3.472	0.756	0.2880	1.203	5.740	0.347	1.653
16	0.750	0.212	0.763	0.9835	1.0168	0.448	1.552	0.440	1.526	3.532	0.750	0.2831	1.282	5.782	0.363	1.637
17	0.728	0.203	0.739	0.9845	1.0157	0.466	1.534	0.458	1.511	3.588	0.744	0.2787	1.356	5.820	0.378	1.622
18	0.707	0.194	0.718	0.9854	1.0148	0.482	1.518	0.475	1.496	3.640	0.739	0.2747	1.424	5.856	0.391	1.609
19	0.688	0.187	0.698	0.9862	1.0140	0.497	1.503	0.490	1.483	3.689	0.733	0.2711	1.489	5.889	0.404	1.596
20	0.671	0.180	0.680	0.9869	1.0132	0.510	1.490	0.504	1.470	3.735	0.729	0.2677	1.549	5.921	0.415	1.585
21	0.655	0.173	0.663	0.9876	1.0126	0.523	1.477	0.516	1.459	3.778	0.724	0.2647	1.606	5.951	0.425	1.575
22	0.640	0.167	0.647	0.9882	1.0120	0.534	1.466	0.528	1.448	3.819	0.720	0.2618	1.660	5.979	0.435	1.565
23	0.626	0.162	0.633	0.9887	1.0114	0.545	1.455	0.539	1.438	3.858	0.716	0.2592	1.711	6.006	0.443	1.557
24	0.612	0.157	0.619	0.9892	1.0109	0.555	1.445	0.549	1.429	3.895	0.712	0.2567	1.759	6.032	0.452	1.548
25	0.600	0.153	0.606	0.9896	1.0105	0.565	1.435	0.559	1.420	3.931	0.708	0.2544	1.805	6.056	0.459	1.541

Appendix B: Simulation and Modeling

Testing and demonstration of the methods for nonnormal data requires data from either real-world processes or computer simulations. The latter avoid any possible problems with the use of proprietary or confidential information, and this appendix shows how to perform them.

Software such as Minitab can generate random nonnormal data directly. Microsoft Excel can generate random uniform, normal (and therefore lognormal), binomial, and Poisson data directly. Select the menu bar for **Tools > Data Analysis > Random Number Generation**.

The simplest way to generate random numbers from other distributions is to use the uniform distribution. Most software packages, including MathCAD and Excel, can return a random number from the range [0,1] with an equal chance of getting any number from this range. Let this number be p. Then find x such that $F(x) = p$, where F is the cumulative probability of the distribution under consideration.

Simulation of the Gamma Distribution

As an example, the following simulation will generate 50 samples of 4 measurements from a gamma distribution with $\alpha = 2$, $\gamma = 2$, and $\delta = 3$. The first step is to generate 50 rows and 4 columns of random numbers from the uniform distribution as shown in Figure B.1. Number of Variables is the number of columns, and Number of Random Numbers is the number of rows. An optional random seed may be added to return the same set of random numbers each time.

Then use the GAMMAINV function on each uniform variable u to return the pth quantile of the gamma distribution. The syntax is GAMMAINV(probability, alpha, beta) where beta is the *reciprocal* of γ. Then add 3 to each result to account for the threshold parameter. As an example, if a random uniform value u is in cell A6, =GAMMAINV(A6,2,0.5)+3 will return the corresponding random number from the gamma distribution with $\alpha = 2$, $\gamma = 2$, and $\delta = 3$.

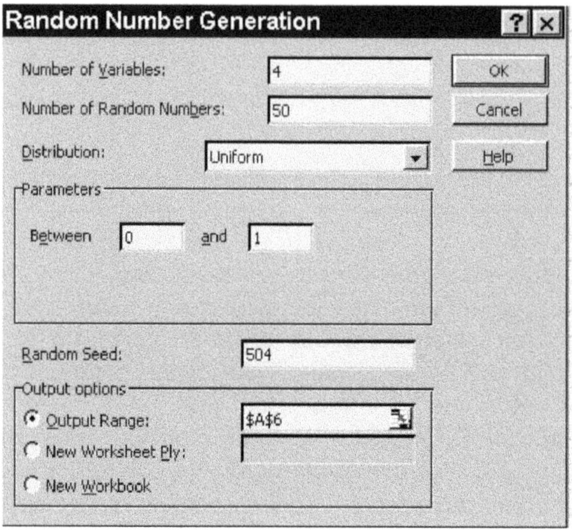

FIGURE B.1
Generation of 50 rows and 4 columns of random uniform variables.

Simulation of the Weibull Distribution

$F(x) = 1 - \exp(-(\frac{x-\delta}{\theta})^\beta) \Rightarrow 1 - F(x) = \exp(-(\frac{x-\delta}{\theta})^\beta)$, and a random uniform number from the interval [0,1] can simulate either $F(x)$ or $1 - F(x)$. $1 - F(x)$ would be the reliability or survivor function in a reliability application. Let u be the random number from the uniform distribution. Then

$$u = \exp\left(-\left(\frac{x-\delta}{\theta}\right)^\beta\right) \Rightarrow -\ln u = \left(\frac{x-\delta}{\theta}\right)^\beta \Rightarrow x = \delta + \theta\left(-\ln u\right)^{1/\beta}$$

Simulation of the Multivariate Normal Distribution

Per Johnson and Wichern (1992, 140–141), let X be a vector of p quality characteristics from a multivariate normal distribution with mean vector μ and covariance matrix Σ. Then $Z = A(X - \mu)$ where Z is a vector of standard normal deviates.

$$A_{ij} = \frac{1}{\sqrt{\lambda_i}} e_{ij}^T$$

where λ_i is the ith eigenvalue of the covariance matrix, and e_{ij} is the ith row and jth column of the eigenvector matrix as shown below. Given a vector of random standard normal deviates Z, rearrangement yields a random multivariate normal vector $X = A^{-1}Z + \mu$.

$$\text{eigenvectors } e = \begin{bmatrix} e_{11} & e_{12} & \cdots & e_{1p} \\ e_{21} & e_{22} & \cdots & e_{2p} \\ \vdots & \vdots & \vdots & \vdots \\ e_{p1} & e_{p2} & \cdots & e_{pp} \end{bmatrix} \quad \text{eigenvalues } \lambda = \begin{bmatrix} \lambda_1 \\ \lambda_2 \\ \vdots \\ \lambda_p \end{bmatrix}$$

Consider the example

$$\Sigma = \begin{bmatrix} 3 & 1 \\ 1 & 2 \end{bmatrix} \quad \text{and } \mu = \begin{bmatrix} 3 \\ 1 \end{bmatrix}$$

The first step is to find the eigenvalues, which requires solution of the equation $|\Sigma - \lambda I| = 0$.[1] The vertical lines indicate the determinant of the matrix, and I is the identity matrix whose diagonal elements are 1 and whose off-diagonal elements are all zero. In this case,

$$\begin{vmatrix} 3-\lambda & 1 \\ 1 & 2-\lambda \end{vmatrix} = (3-\lambda)(2-\lambda) - 1 = \lambda^2 - 5\lambda + 5 = 0$$

which has the solutions 3.618 and 1.382. This becomes more complicated as the size of the matrix increases, but MathCAD's eigenvals and eigenvecs functions will perform the necessary calculations automatically. Microsoft Excel does not have routines for the eigenvalues and eigenvectors, but it may be possible to use the built-in matrix determinant function and the Goal Seek capability to improvise a solution.[2]

Next, solve $\Sigma x = \lambda x$ for each eigenvalue.

$$\begin{bmatrix} 3 & 1 \\ 1 & 2 \end{bmatrix} \begin{bmatrix} x_{11} \\ x_{21} \end{bmatrix} = 3.618 \begin{bmatrix} x_{11} \\ x_{21} \end{bmatrix} \Rightarrow \begin{bmatrix} 3x_{11} + x_{21} \\ x_{11} + 2x_{21} \end{bmatrix} = \begin{bmatrix} 3.618x_{11} \\ 3.618x_{21} \end{bmatrix} \Rightarrow x_{11} = 1.618x_{21}$$

$$\begin{bmatrix} 3 & 1 \\ 1 & 2 \end{bmatrix} \begin{bmatrix} x_{12} \\ x_{22} \end{bmatrix} = 1.382 \begin{bmatrix} x_{12} \\ x_{22} \end{bmatrix} \Rightarrow \begin{bmatrix} 3x_{12} + x_{22} \\ x_{12} + 2x_{22} \end{bmatrix} = \begin{bmatrix} 1.382x_{12} \\ 1.382x_{22} \end{bmatrix} \Rightarrow x_{22} = -1.618x_{12}$$

This does not seem particularly useful because there are infinite solutions to these equations, but there is an additional condition that each eigenvector has a length of 1. Define each x arbitrarily, e.g., with a base value of 1, and then the eigenvector is

$$e = \frac{x}{\sqrt{x^T x}}$$

In this case, let x_{21} equal 1 and x_{12} equal -1.

For $\lambda = 3.618$,

$$x = \begin{bmatrix} 1.618 \\ 1 \end{bmatrix} \Rightarrow x^T x = \begin{bmatrix} 1.618 & 1 \end{bmatrix} \begin{bmatrix} 1.618 \\ 1 \end{bmatrix} = 3.618 \text{ and } e = \begin{bmatrix} \dfrac{1.618}{\sqrt{3.618}} \\ \dfrac{1}{\sqrt{3.618}} \end{bmatrix}$$

For $\lambda = 1.382$,

$$x = \begin{bmatrix} -1 \\ 1.618 \end{bmatrix} \Rightarrow x^T x = \begin{bmatrix} -1 & 1.618 \end{bmatrix} \begin{bmatrix} -1 \\ 1.618 \end{bmatrix} = 3.618 \text{ and } e = \begin{bmatrix} \dfrac{-1}{\sqrt{3.618}} \\ \dfrac{1.618}{\sqrt{3.618}} \end{bmatrix}$$

and finally

$$e = \begin{bmatrix} 0.851 & -0.526 \\ 0.526 & 0.851 \end{bmatrix}$$

As a check, the sum of the squares of 0.851 and 0.526 equals 1. As another check, the spectral decomposition

$$\sum_{j}^{p} \lambda_i e_{.j} e_{.j}^T = \Sigma$$

where e_j is the jth column of the eigenvalue matrix returns the original covariance matrix.

$$3.618 \begin{bmatrix} 0.851 \\ 0.526 \end{bmatrix} [0.851 \quad 0.526] + 1.382 \begin{bmatrix} -0.526 \\ 0.851 \end{bmatrix} [-0.526 \quad 0.851]$$

$$= 3.618 \begin{bmatrix} 0.724 & 0.448 \\ 0.448 & 0.277 \end{bmatrix} + 1.382 \begin{bmatrix} 0.277 & -0.448 \\ -0.448 & 0.724 \end{bmatrix} = \begin{bmatrix} 3 & 1 \\ 1 & 2 \end{bmatrix}$$

Now compute matrix A for the simulation.

$$e^T = \begin{bmatrix} 0.851 & 0.526 \\ -0.526 & 0.851 \end{bmatrix} \quad \text{and then} \quad A_{ij} = \frac{1}{\sqrt{\lambda_i}} e_{ij}^T$$

$$\Rightarrow A = \begin{bmatrix} \dfrac{0.851}{\sqrt{3.618}} & \dfrac{0.526}{\sqrt{3.618}} \\ \dfrac{-0.526}{\sqrt{1.382}} & \dfrac{0.851}{\sqrt{1.381}} \end{bmatrix} = \begin{bmatrix} 0.447 & 0.277 \\ -0.277 & 0.724 \end{bmatrix}$$

This is the basis of the simulated data in sheet Multivariate_1 of the simulated data spreadsheet.

Endnotes

1. Johnson and Wichern, 1992, 84–85.
2. http://controls.engin.umich.edu/wiki/index.php/EigenvaluesEigenvectors#Microsoft_Excel

Appendix C: Numerical Methods

Bisection

Bisection is a highly robust algorithm that will always find the root of an equation $f(x)$ provided that the solution (1) is unique and (2) lies within the interval $[a,b]$. Begin by setting x_L and x_R equal to a and b, respectively. Then compute the function and (Hornbeck, 1975, 66):

1. If $f(x_L) \times f(x_R) < 0$, then go to step 3, otherwise go to step 2.
   - If the functions of x_L and x_R have opposite signs, the root is between x_L and x_R. If not, the root is outside this interval.
   - This condition will always be true in the first iteration unless the root is not in the interval $[a,b]$, so the *TEMP* variable shown below will always be defined. It is useful to perform this test up front to make sure the answer is in fact somewhere between a and b.

2. The root is outside the interval $[x_L, x_R]$: Let $x_R = x_L$, $x_L = TEMP$, and then

$$x_L = \frac{x_L + x_R}{2}.$$

Return to step 1.

3. The root is inside the interval $[x_L, x_R]$: If $x_R - x_L < 2\varepsilon$ where ε is the desired tolerance for the result, then the solution is

$$\frac{x_L + x_R}{2}$$

(stopping rule). Otherwise:

   - Let $TEMP = x_L$, and then

$$x_L = \frac{x_L + x_R}{2},$$

and then return to step 1.

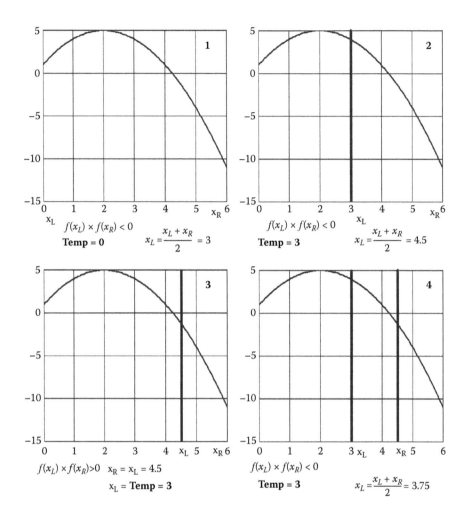

FIGURE C.1
First four iterations of a bisection.

- Note that, in both cases,

$$x_L = \frac{x_L + x_R}{2}.$$

Figure C.1 shows how this procedure converges on the root of $f(x) = 1 + 4x - x^2$. The first two iterations apply the rule for $f(x_L) \times f(x_R) < 0$, which increases x_L by half the distance to x_R. In the third iteration, $f(x_L) \times f(x_R) > 0$ means the root is outside the interval, so x_L moves back to its previous position while x_R moves left by half the distance to the new x_L. The fourth iteration shows how the interval will continue to converge on the solution.

Response Surface Optimization

Given the need to optimize two parameters whose limits are $[a_1,b_1]$ and $[a_2,b_2]$, respectively, create a box with these as the corner points and

$$\left(\frac{a_1+b_1}{2}, \frac{a_2+b_2}{2} \right)$$

as the center point. *Note, however, that if the solution is outside this starting box, the algorithm will nonetheless find it under most circumstances.* The box will move toward the solution even if the solution is outside the boundaries that the user defines initially. This differs from bisection because the bisection cannot find the answer if it is outside the starting interval $[a,b]$. Then:

$$\text{Lower left} \quad LL = (a_1, b_1)$$
$$\text{Lower right} \quad LR = (a_2, b_1)$$
$$\text{Upper left} \quad UL = (a_1, b_2)$$
$$\text{Upper right} \quad UR = (a_2, b_2)$$
$$\Delta x = \frac{a_2 - a_1}{2} \quad \text{and} \quad \Delta y = \frac{b_2 - b_1}{2}$$
$$\text{Center} \quad CTR = (a_1 + \Delta x, b_1 + \Delta y) = \left(\frac{a_1 + a_2}{2}, \frac{b_1 + b_2}{2} \right)$$

Calculate the objective function $f(x,y)$ for each of the five points. The best result is the largest if the objective is maximization, the smallest if the objective is minimization, and the smallest absolute value or square if the objective is the root at which $f(x,y) = 0$. (Only the first application has been tested in this book.) Then:

1. If a corner position contains the best result, it becomes the center point for the next iteration. This causes the box or rectangle to seek the top of the response surface.
   - If the corner position in question is (x,y),

$$CTR = (x, y)$$
$$LL = (x - \Delta x, y - \Delta y)$$
$$LR = (x + \Delta x, y - \Delta y)$$
$$UL = (x - \Delta x, y + \Delta y)$$
$$UR = (x + \Delta x, y + \Delta y)$$

- If there is a constraint on the solution, such as the fact that a parameter cannot be less than zero, make sure none of the new corner positions violates this constraint.

2. If the center position contains the best result, the optimum lies somewhere inside the rectangle. Contract the square by 50 percent on each side and repeat the calculations for the new corner squares.

- $\Delta x = \frac{\Delta x}{2}$ and $\Delta y = \frac{\Delta y}{2}$. Then, given CTR = (x,y),

$$LL = (x - \Delta x, y - \Delta y)$$

$$LR = (x + \Delta x, y - \Delta y)$$

$$UL = (x - \Delta x, y + \Delta y)$$

$$UR = (x + \Delta x, y + \Delta y)$$

- If this reduces Δx or Δy to less than the desired tolerance (noting that the stopping rule requires both of them to be less than the tolerance), it is optional to stop the contraction in the indicated direction.

3 Continue until the square contracts to the desired tolerance for the solution.

A one-dimensional version is available as an alternative to bisection, and it was necessary to use it to get the confidence limits for the survivor function (or nonconforming fraction) of the gamma distribution. In this case, define an interval $[x_L, x_R]$, although the solution need not lie in this interval as it does for bisection. There must, however, be a unique maximum or minimum and no local minima or maxima at which the algorithm can settle. Then:

$$\text{Left} = x_L$$

$$\text{Center} = \frac{x_L + x_R}{2}$$

$$\text{Right} = x_R$$

$$\Delta = \frac{x_R - x_L}{2}$$

Calculate the objective function for each of the three positions. If an end point yields the best result, it becomes the new center point and then the new end points are the center plus or minus Δ. If the center point yields the best results, contract the interval by 50 percent. Divide Δ by 2, and then the end points are the center point plus or minus Δ.

References

American Society for Testing and Materials (ASTM). 1990. *Manual on Presentation of Data and Control Chart Analysis*, 6th ed. Philadelphia, PA: American Society for Testing and Materials.

Arnold, Horace Lucien, and Fay Leone Faurote. 1915. Ford methods and the Ford shops. *The Engineering Magazine*. Reprinted 1998, North Stratford, NH: Ayer Company Publishers, Inc.

AT&T. 1985. *Statistical quality control handbook*. Indianapolis, IN: AT&T Technologies.

Automotive Industry Action Group (AIAG). 2002. *Measurement systems analysis*, 3rd ed. Available at http://www.aiag.org.

Automotive Industry Action Group (AIAG). 2005. *Statistical Process Control*, 2nd ed. Available at http://www.aiag.org.

Barrentine, Larry B. 1991. *Concepts for R&R studies*. Milwaukee, WI: ASQ Quality Press.

Beyer, William H. 1991. *Standard probability and statistics tables and formulae*. Boca Raton, FL: CRC Press.

Box, George, and Alberto Luceño. 1997. *Statistical control by monitoring and feedback adjustment*. New York: John Wiley & Sons.

Burke, R. J., R. D. Davis, and F. C. Kaminsky. 1991. Resorting to statistical terrorism in quality control. 47th Annual Conference of American Society for Quality, Rochester Section.

Chou, Youn-Min, D. B. Owen, and Salvador A. Borrego. 1990. Lower confidence limits on process capability indices. *Journal of Quality Technology* 22(3): July 1990.

Chou, Youn-Min, Alan Polansky, and Robert L. Mason. 1998. Transforming non-normal data to normality in statistical process control. *Journal of Quality Technology* 30 (2): 133–141.

Cooper, P. J., and N. Demos. 1991. Losses cut 76 percent with control chart system, *Quality*, April.

Crofts, Alfred E. 1982. On a property of the F distribution. *Trabajos de Estadística y de Investigacion Operativa* 33 (2): 110-111.

Davis, Robert, and Frank Kaminsky. 1990. Process capability analysis and Six Sigma. Mid-Hudson Valley (New York) section of the American Society of Quality dinner meeting, 30 May.

Ford, Henry, and Samuel Crowther. 1926. *Today and tomorrow*. New York: Doubleday, Page & Company. (Reprint available from Productivity Press, 1988.)

Ford, Henry, and Samuel Crowther. 1930. *Moving forward*. New York: Doubleday, Doran, & Company.

Goldratt, Eliyahu, and Cox, Jeff. 1992. *The Goal*. Croton-on-Hudson, NY: North River Press.

Guttman, Irwin, S. S. Wilkes, and J. Stuart Hunter. 1982. *Introductory Engineering Statistics*. New York: John Wiley & Sons.

Heinlein, Robert A. *Starship Troopers*. 1959. New York: Putram.

Holmes, Donald S., and A. Erhan Mergen. 1993. Improving the performance of the T^2 control chart. *Quality Engineering* 5 (4): 619–625.

Holmes, Donald S., and A. Erhan Mergen. 1998. A multivariate test for randomness. *Quality Engineering* 10 (3): 505–508.

Holmes, Donald S., and A. Erhan Mergen. 1998a. Measuring process performance for multiple variables. *Quality Engineering* 11 (1): 55–59.

Holmes, Donald S., and A. Erhan Mergen. 2001. Measuring process performance for multiple variables: Revisited. *Quality Engineering* 13 (4): 661–665.

Holmes, Donald S., and A. Erhan Mergen. 2004. A multivariate subgroup rationality test. *Quality Engineering* 16 (4): 657–662.

Hornbeck, R. W. 1975. *Numerical methods.* New York: Quantum Publishers.

Hradesky, John. 1988. *Productivity and quality improvement: A practical guide to implementing statistical process control.* New York: McGraw-Hill

Hradesky, John. 1995. *Total quality management handbook.* New York: McGraw-Hill.

Jacobs, David C. 1990. Watch out for nonnormal distributions. *Chemical & Engineering Progress* November: 19–27.

Johnson, Norman, and Samuel Kotz. 1970. *Distributions in statistics: Continuous univariate distributions-1.* Boston, MA: Houghton-Mifflin.

Johnson, Richard A., and Dean W. Wichern. 1992. *Applied multivariate statistical analysis,* 3rd ed. Englewood Cliffs, NJ: Prentice Hall.

Juran, Joseph, and Frank Gryna. 1988. *Juran's quality control handbook,* 4th ed. New York: McGraw-Hill.

Kapur, K. C., and L. R. Lamberson. 1977. *Reliability in Engineering Design.* New York: John Wiley & Sons.

Kramer, C. Y., and D. R. Jensen. 1969. Fundamentals of multivariate analysis—Part II: Inference about two treatments. *Journal of Quality Technology* 1 (3): 189–204.

Kushler, R. H., and P. Hurley. 1992. Confidence bounds for capability indices. *Journal of Quality Technology* 24 (4): 188–195.

Lawless, J. F. 1982. *Statistical models and methods for lifetime data.* New York: John Wiley & Sons.

Levinson, W. A. 1994. Statistical process control in microelectronics manufacturing. *Semiconductor International,* November.

Levinson, W. A. 1994a. Multiple attribute control charts, *Quality,* December, 10–13.

Levinson, W. A. 1995. How good is your gage? *Semiconductor International* October: 165–169.

Levinson, W. A. 1996. Do you need a new gage? *Semiconductor International* February: 113–118.

Levinson, W. A. 1997a. Exact confidence limits for process capabilities. *Quality Engineering* 9 (3): 521–528.

Levinson, William A. 1998a. Using SPC in batch processes. *Quality Digest,* March: 45–48.

Levinson, William, and Angela Polny. 1999. SPC for tool particle counts. *Semiconductor International,* June.

Levinson, William. 1999a. Statistical characterization of a die shear test. *Quality Engineering* 11 (3): 443–448.

Levinson, William A., and Joseph Cumbo. 2000. Simplified control charts for multivariate normal systems. *Quality Engineering,* June

Levinson, W. A., Frank Stensney, Raymond Webb, and Ronald Glahn. 2001. SPC for particle counts. *Semiconductor International,* October.

Levinson, William A., Donald S. Holmes, and A. Erhan Mergen. 2002. Variation charts for multivariate processes. *Quality Engineering* 14 (4): 539–545.

Levinson, William, and Raymond Rerick. 2002. *Lean Enterprise: A Synergistic Approach for Minimizing Waste*. Milwaukee, WI: ASQ Quality Press.

Levinson, William. 2004. Control charts control multiple attributes. *Quality*, September: 40–43.

Lieberman, Gerald J. 1957. Tables for one-sided statistical tolerance limits. Technical Report No. 37, November 1 1957. Applied Mathematics and Statistics Laboratory, Stanford University, prepared for Office of Naval Research.

Mason, Robert L., Youn-Min Chou, and John C. Young. 2001. Applying Hotelling's T^2 statistic to batch processes. *Journal of Quality Technology* 33 (4):

Mason, Robert L., Nola D. Tracy, and John C. Young. 1995. Decomposition of T^2 for multivariate control chart interpretation. *Journal of Quality Technology* 27 (2): 99–108.

Mason, Robert L., Nola D. Tracy, and John C. Young. 1997. A practical approach for interpreting multivariate T^2 control chart signals. *Journal of Quality Technology* 29 (4).

Mason, Robert L., and John C. Young. 2005. Multivariate tools: Principal component analysis. *Quality Progress*, February: 83–85.

Messina, William. 1987. *Statistical quality control for manufacturing managers*. New York: John Wiley & Sons.

Mood, Alexander M., Franklin A. Graybill, and Duane C. Boes. 1974. *Introduction to the theory of statistics*, 3rd ed. New York: McGraw-Hill.

Montgomery, 1991. *Introduction to statistical quality control*, 2nd ed. New York: John Wiley & Sons.

Page, Michael. 1994. Analysis of non-normal process distributions. *Semiconductor International*, October: 88–96.

Papp, Zoltan. 1992. Two-sided tolerance limits and confidence bounds for normal populations. *Commun. Statitist. Theory Methods* 21 (5): 1309–1318.

Sauers, Dale G. 1997. Hotelling's T^2 statistic for multivariate statistical process control: A nonrigorous approach. *Quality Engineering* 9 (4): 627–634.

Shapiro, Samuel S. 1986. *How to test normality and other distributional assumptions*. Milwaukee, WI: ASQ Quality Press.

Vargas, N. Jose, and C. Juliana Lagos. 2007. Comparison of multivariate control charts for process dispersion. *Quality Engineering* 19, 191–196.

Wald, A., and J. Wolfowitz. 1946. Tolerance limits for a normal distribution. *Annals of. Math. Statist.* 17 (2): 208–215.

Wang, F. K., and James C. Chen. 1998. Capability analysis using principal components analysis. *Quality Engineering* 11 (1): 21–27.

Wasserman, Gary S. 2000. Easy ML estimation of normal and Weibull metrics. *Quality Engineering* 12 (4): 569–581.

Wilks, S. S. 1948. Order statistics. *Bull. Amer. Math. Soc.* 54 (Part 1): 6–50.

Wortman, Bill. 1991. *Certified Six Sigma Black Belt primer*. West Terre Haute, IN: Quality Council of Indiana.

Index

For Product Safety Concerns and Information please contact our EU
representative GPSR@taylorandfrancis.com
Taylor & Francis Verlag GmbH, Kaufingerstraße 24, 80331 München, Germany